CHEMICAL ENGINEERING METHODS AND TECHNOLOGY

BIOMEDICAL CHROMATOGRAPHY

CHEMICAL ENGINEERING METHODS AND TECHNOLOGY

Additional books in this series can be found on Nova's website under the Series tab.

Additional E-books in this series can be found on Nova's website under the E-book tab.

CHEMICAL ENGINEERING METHODS AND TECHNOLOGY

BIOMEDICAL CHROMATOGRAPHY

JOHN T. ELWOOD
EDITOR

Nova Science Publishers, Inc.
New York

For permission to use material from this book please contact us:
Telephone 631-231-7269; Fax 631-231-8175
Web Site: http://www.novapublishers.com

LIBRARY OF CONGRESS CATALOGING-IN-PUBLICATION DATA

Biomedical chromatography / editor, John T. Elwood.
p. ; cm.
Includes bibliographical references and index.
ISBN 978-1-60741-291-5 (hardcover)
1. Drugs--Analysis. 2. Chromatographic analysis. 3. Pharmaceutical chemistry. I. Elwood, John T.
[DNLM: 1. Pharmaceutical Preparations--chemistry. 2. Pharmacokinetics. 3. Chromatography--methods. QV 38 B6155 2010]
RS189.5.C48B56 2010
615'.1901--dc22

2010001737

Published by Nova Science Publishers, Inc. ✦ New York

CONTENTS

PREFACE

Biomedical chromatography is the science devoted to the application of chromatography and allied techniques in the biological and medical sciences. Under research are the methods and techniques relevant to the separation, identification and determination of substances in biochemistry, biotechnology, molecular biology, cell biology, clinical chemistry, pharmacology and related disciplines. These include the analysis of body fluids, cells and tissues, purification of biologically important compounds, and such. This new book discusses and reviews these and related topics.

Chapter 1 - The discovery and development of new drugs is an arduous and complex process in which bioanalytical progress has emerged as a driving force of the utmost importance. Over the last years, many quantitative chromatographic assays have been continuously developed for new chemical entities and their main metabolites to support the various stages of drug discovery and preclinical and clinical development. However, the bioanalytical methods originally developed during the phase of drug discovery may not be suitable to support further drug development, where the availability of bioanalytical methods fully validated will be required to obtain reliable analytical data. Bearing in mind that the pharmacokinetics is the branch of pharmacology dedicated to the assessment of the absorption, distribution, metabolism, and excretion (ADME) of drugs by the body, it is evident that the pharmacokinetic studies necessarily involve the development of quantitative bioanalytical methods for the measurement of drugs and their metabolites in biological samples of interest. Nowadays, pharmacokinetic properties are considered on the heels of rational drug design and, as soon as possible, at the earliest phases of drug discovery, new molecules are screened in relation to their ADME profiling using predictive in vitro models. Then, at the preclinical stage of drug development, a lot of pharmacokinetic studies are also performed in several animal species to demonstrate the therapeutic potential of an investigational new molecule and to ensure that it will be safe to entry into man. Nevertheless, preclinical testing, mainly in whole-animal models, follows closely all clinical phases of drug development, since for ethical and practical reasons some useful experiments cannot be performed in human.

In this chapter, the authors intend to discuss the bioanalytical framework in drug discovery and development process and to discuss, with emphasis on liquid chromatography, the bioanalytical method validation and its use in routine analysis to support pharmacokinetic studies.

Chapter 2 - The study of the interaction of drugs with plasmatic proteins is essential for the understanding of the pharmacological properties of drugs. At the pharmacokinetic level, bound drug to proteins is a reserve of drug in the organism since it cannot be distributed, metabolized and excreted; so, only free drug is active since it can cross barriers and move towards tissues. The interaction of drugs with proteins can be of different types: hydrophobic, electronic or steric. In this chapter, the methodologies employed for characterizing these interactions are revised and compared. The interactions of β-blockers, phenothiazines, antihistamines, flavonoids, benzodiazepines and barbiturates with human serum albumin (HSA), α1-glycoprotein acid (AGP) and lipoproteins (LIPO) and/or globulins (GLOB) are evaluated by capillary electrophoresis and summarized. The use of chemometrics to characterize the nature of interactions is also reviewed.

Chapter 3 - Bioequivalence studies play a major role in the development of new drugs and in the marketing of generic formulations; such trials also contribute in to access to low-cost and effective medicines in developing countries. At present, with the loss of patents of novel molecules, the difficulty in designing interchangeability trials has increased.

Several factors contribute to the complexity of bioequivalence studies. Some new drugs are more potent, and their concentrations in biological fluids are lower each time. In addition to this, the task of controlling the content uniformity of the formulations of these drugs is more demanding

Finally, due to the global politics of generic usage, it has become evident that there is minimal information concerning metabolic background in different populations, which has comprised another unexpected source of variation.

Chapter 4 - Pharmaceuticals are “emerging contaminants” in the environment, which have received increasing attention in recent years, in part, because these compounds are constantly being emitted into the environment in quantities similar to other organic contaminants, and thus may be candidates for future regulation. This paper reviews recently published data with respect to sample-preparation techniques for determining pharmaceutical traces in various different environmental matrices using liquid chromatography. An overview of multi-residue analytical methods, covering sample extraction and purification as well as chromatographic separation and various different detection methods is provided to illustrate common trends and method variability. Additionally, recent developments and advances in this field are presented.

Chapter 5 - Peroxyoxalate chemiluminescence (POCL) reaction is known as a useful, selective and sensitive detection system especially when coupled with separation techniques such as high-performance liquid chromatography (HPLC) or capillary electrophoresis. The authors have developed analytical methods for catecholamines and/or their metabolites using HPLC-POCL reaction detection. This article reviews 1) the analytical methods for catecholamines, 2) the analytical methods for catecholamines and their 3-*O*-methyl metabolites, and 3) the application of these methods to bio-samples for the clarification of the role of catecholamines metabolism in blood pressure regulation.

Chapter 6 - A sample of 5 mM naproxen in methanol or ethanol was photoirradiated with a Hanovia 200 W high-pressure quartz Hg lamp. In total, 6 photoproducts derived from each sample were observed from the HPLC chromatograms. Four major photoproducts were separated, and their structures elucidated by EI-MS and various spectroscopic methods. A reaction scheme of naproxen in alcoholic solvents is proposed: the photochemical reaction routes occur mainly via decarboxylation and esterification, followed by oxidation with singlet oxygen to produce an alcohol and a ketone.

In: Biomedical Chromatography
Editors: John T. Elwood, pp.1-28
ISBN: 978-1-60741-291-5

Chapter 1

BIOANALYSIS IN DRUG DISCOVERY AND DEVELOPMENT: FOCUS ON PHARMACOKINETIC STUDIES

Gilberto Alves, Ana Fortuna and Amílcar Falcão*
Laboratory of Pharmacology, Faculty of Pharmacy & Center for Neurosciences and Cell Biology, Coimbra University, Coimbra, Portugal

ABSTRACT

The discovery and development of new drugs is an arduous and complex process in which bioanalytical progress has emerged as a driving force of the utmost importance. Over the last years, many quantitative chromatographic assays have been continuously developed for new chemical entities and their main metabolites to support the various stages of drug discovery and preclinical and clinical development. However, the bioanalytical methods originally developed during the phase of drug discovery may not be suitable to support further drug development, where the availability of bioanalytical methods fully validated will be required to obtain reliable analytical data. Bearing in mind that the pharmacokinetics is the branch of pharmacology dedicated to the assessment of the absorption, distribution, metabolism, and excretion

* Corresponding author: E-mail address: acfalcao@ff.uc.pt, Laboratory of Pharmacology, Faculty of Pharmacy, Coimbra University, 3000-141 Coimbra, Portugal, Phone: +351-239-855089/90, Fax: +351-239-855099.

(ADME) of drugs by the body, it is evident that the pharmacokinetic studies necessarily involve the development of quantitative bioanalytical methods for the measurement of drugs and their metabolites in biological samples of interest. Nowadays, pharmacokinetic properties are considered on the heels of rational drug design and, as soon as possible, at the earliest phases of drug discovery, new molecules are screened in relation to their ADME profiling using predictive in vitro models. Then, at the preclinical stage of drug development, a lot of pharmacokinetic studies are also performed in several animal species to demonstrate the therapeutic potential of an investigational new molecule and to ensure that it will be safe to entry into man. Nevertheless, preclinical testing, mainly in whole-animal models, follows closely all clinical phases of drug development, since for ethical and practical reasons some useful experiments cannot be performed in human.

In this chapter, we intend to discuss the bioanalytical framework in drug discovery and development process and to discuss, with emphasis on liquid chromatography, the bioanalytical method validation and its use in routine analysis to support pharmacokinetic studies.

Key words: Drug discovery and development; Bioanalysis; Method validation; Pharmacokinetics.

INTRODUCTION

Over the years the pharmaceutical research has focused its attention mainly in the discovery and development of new drugs, but this goal has not been easily attained. Indeed, the drug discovery and development (DDD) is a challenging, expensive and time-consuming process and it requires the contribution of multidisciplinary teams of scientists, working together, to identify disease targets and to find chemical compounds with pharmacological activity and desirable drug-like properties. Nowadays, in spite of the increasing emergency of biological drugs such as antibodies, the identification of small bioactive synthetic molecules remains the major approach to bring new drugs to the market. This strategy is essentially dependent of the high-throughput screening (HTS) of new chemical entities (NCEs) obtained from chemical libraries generated by combinatorial chemistry. These methodologies of high-throughput molecular screening are valuable tools to select drug candidates with potentially improved pharmacological profiles. Of thousands of compounds that usually are screened in a program of DDD, only a small fraction of these will have suitable pharmacodynamic and pharmacokinetic properties to become a drug. Hence, there is no doubt that chemistry, biology and pharmacology disciplines play a central role in the discovery and development of novel therapeutic agents. Likewise, no less important appears to be the bioanalysis, which support the different phases of drug discovery and preclinical and clinical development [1-4].

OVERVIEW OF THE DRUG DISCOVERY AND DEVELOPMENT PROCESS

The process of DDD has undergone an amazing evolution throughout the time. The earliest approach of drug discovery was the use of natural products and it is known as long as history is recorded. Much later, this lead to the screening of natural and synthetic compounds in order to maximize in vitro and in vivo potency and selectivity against relevant biological targets [5-7]. In fact, until a few years ago, the pharmaceutical companies used the traditional linear approach to develop new drugs, where drug discovery and drug development departments were independents. The first one was in charge for the synthesis of small quantities of NCEs and for their efficacy evaluation against the disease process, whereas the second one was responsible for absorption, distribution, metabolism and excretion (ADME) prediction, and pharmacokinetic and toxicological characterization. However, this DDD approach was not as effective as expected because the number of NCEs was enhancing but not the number of new drugs introduced in the market. Indeed, almost 40% of compounds that entered in clinical phases did not attain the market due to their undesirable and poor pharmacokinetic properties [8-12]. For that reason, pharmaceutical companies needed to reorganize from traditional linear drug development approach to the parallel one, where the pharmacological efficacy is screened all together with ADME properties from the earliest stages of drug discovery. Hence, the process of DDD presently involves the target identification, hit identification, lead identification and optimization, preclinical development and clinical development (Figure 1). First of all, the rational drug design requires the knowledge of the disease state being targeted, as well as the understanding of its inherent biochemical and molecular mechanisms. NCEs are then evaluated through a variety of high-throughput in silico or bioinformatics assays to identify the drug candidates with higher likelihood of success. This phase, called hit identification, is where large compound libraries are screened in a relative short period of time, attempting to find the molecules with higher potential to become a novel drug. It is common to screen chemical libraries as large as 100.000 compounds in a few months using HTS technologies. Given that the development of all hits identified would be too expensive and would take too much time, it is necessary to select a main drug candidate and, if desired, some backup candidates – lead identification stage. The next step includes the lead optimization, which consists of an iterative process of synthesis and testing, being the experimental data correlated with the corresponding chemical structures to find the most promising leads. Afterwards, these selected NCEs are promoted from the drug discovery to preclinical development phase, implying pharmacokinetic, pharmacodynamic, and toxicological studies in a number of animal species. NCEs that are successful in all referred stages may enter into man and start their clinical development [9-12]. This strategy has reduced the pharmacokinetic attrition rate to nearly 10%, probably due to the increased endeavour in applying pharmacokinetic principles during initial phases of DDD. Actually, the relationship between inadequate ADME properties and failure of drug candidates in clinical development seems to be clear [9,12,13].

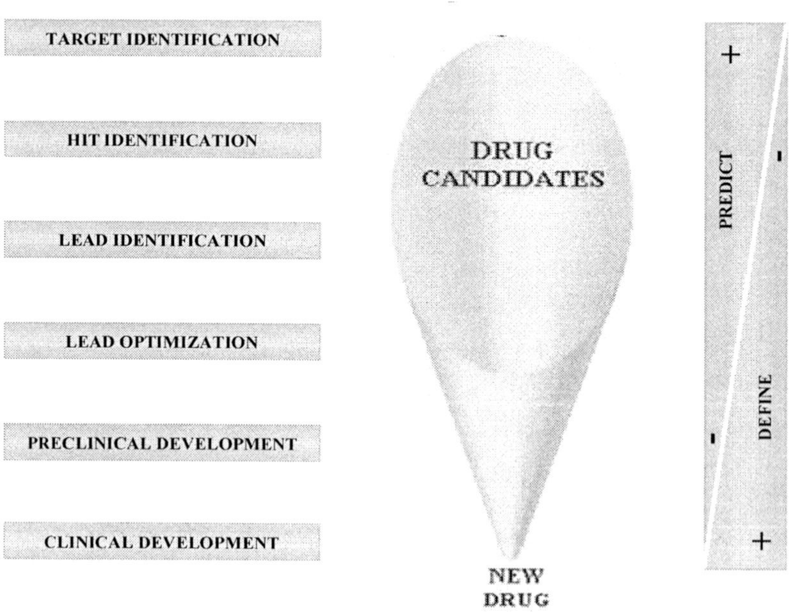

Figure 1. Stages in the discovery and development of new drugs.

The principal goal of early ADME screenning is to reveal compounds with flawed physico-chemical properties earlier, filtering out them and avoiding their passage into more expensive and time-consuming phases. Moreover, early ADME evaluation also informs about fundamental mechanisms and specific properties that may be used to design and select the best compounds. All this information, if obtained earlier, can help the chemists to modify specific functionalities of a molecule improving the drug design. During the phase of drug discovery the major pharmacokinetic properties to be predicted for NCEs are in vivo absorption, plasma protein binding, first-pass metabolism, competition among metabolic enzymes and intrinsic clearance. For instance, the parallel artificial membrane permeability assay (PAMPA), developed by Kansy et al. [14], is a rapid, cost-effective and high-throughput in vitro assay to estimate the potential absorption of new compounds. It is used to assess passive transcellular drug permeability across a phospholipidic system that mimics biological membranes such as the gastrointestinal mucous membrane [14-20], the blood-brain barrier [21,22], and the skin [23]. This non-cellular assay is valuable at the drug discovery stage as an absorption screening tool wherein the main purpose is to eliminate poorly absorbed compounds [24-26]. The human epithelial colon adenocarcinoma cell line (Caco-2), is also widely used for predicting gastrointestinal drug absorption in humans [27-29]. The ability of these cells to express human intestinal transporters and enzymes allows its use for the study of passive and active transport mechanisms which may be involved in drug absorption [30-34]. The screening of the extent of plasma protein binding of NCEs is also important since it affects their disposition, potency, and safety. Several techniques have been used to support drug-plasma protein binding studies, but equilibrium dialysis and ultrafiltration are the most widespread ones [35-38]. Besides absorption and plasma protein binding, the metabolism is perhaps the key determinant of the fate of a drug in the body, being also evaluated during the initial phase of DDD. Therefore, the metabolic stability of NCEs is usually evaluated in suspensions of liver microsomes and hepatocytes obtained from several animal species and/or humans. NCEs that are rapidly metabolized are shown to be less promising than those relatively stables. If a compound is quickly metabolized the likelihood to originate less active or inactive metabolites is higher, decreasing its therapeutic potential [39,40]. Hence, it is important to identify early in DDD process the main metabolites produced in humans to understand their relevance from a pharmacological and toxicological point of view. Finally, in vitro studies employing human liver microsomes are also crucial to identify the main metabolizing enzymes responsible for the metabolism of each drug candidate and to foresee its potential for drug-drug interactions [41-43]. All these assays may be performed in multi-well plates and easily automated, enhancing the data acquisition rate and allowing a HTS.

The screening of pharmacokinetic properties can also be done resorting to in vivo studies. Actually, the pharmaceutical industry is increasingly using them in the initial states of the process of DDD [44,45]. Nevertheless, these studies must offer enough speed for HTS required at the drug discovery stage. Conventional full pharmacokinetic studies include the intravenous and oral administration of each compound to groups of 6-8 animals and the blood samples are taken at different 6-12 points of time. However, the high number of animals and the labour-intensive sample collection are a hurdle to the HTS of these assays. Hence, in order to overcome this problem considerable efforts have been done to develop new methodologies. For example, the multi-compound dosing, also called cassette or N-in-one dosing, has been widely applied at the initial phases of DDD process to minimize the problem of low-throughput screening, making use of the liquid chromatography coupled with tandem mass spectrometry (LC-MS-MS) [46-51]. This approach implies the administration of multiple NCEs simultaneously to laboratory animals followed by their individual determination, in parallel, at different time-points. Although frequently used due to the small number of animals needed as so as the few amount of samples obtained, this type of assays has some scientific limitations. Major restrictions are related with the potential for interactions between drug candidates, inaccurate pharmacokinetic data and difficulties in preparing a convenient dosing formulation of multiple compounds. These limitations may be overcome by consecutive administration of drug candidates in low doses and in short intervals of time. In these conditions, cassette dosing remains a valuable tool in early stages of DDD and it allows the comparison of similar NCEs based on their corresponding pharmacokinetic parameters. Even though, many authors have adopted other strategies that permit fast data acquisition from individually dosed animals, essentially by shortening the plasma and tissue sampling period to 6-8 h, which is shorter than that in conventional pharmacokinetic studies, and/or by pooling the samples collected at the same post-dosing time-points [45,52,53]. Although the preceding approach overcomes the problems related to drug-drug interactions, one of the major advantages of cassette-dosing is lost: the reduction of animal usage and handling.

The employment of these in vitro and in vivo screening tools at the level of drug discovery will reduce the turnaround time for delivery information to chemists during the lead optimization phase and will support the prioritisation of the screened compounds. At early phases of DDD, the drug candidates are usually classified into one of the following broad categories: clearly bad, marginal or probably good. Those compounds classified as clearly bad are eliminated and the good ones are rank-order for in-depth studies. If enough good candidates are found, it may be unnecessary to make hard decisions about the marginal ones. However, the accuracy of the numerical data obtained in these screening studies is not much important since at this stage only matters the correct prioritisation of drug candidates for further evaluation in more costly in vivo assays. These in vivo studies are performed first in laboratory animals (preclinical development) and then in humans (clinical development). The preclinical drug development, which constitutes the last critical stage before human exposure to drug candidate, involves the detailed characterization of its pharmacokinetics, efficacy, and safety in several rodent and non-rodent species [54]. Oral bioavailability, area under the concentration-time curve, half-life, clearance, volume of distribution, and mean residence time are the typically pharmacokinetic parameters determined from plasma concentration-time profiles after oral and intravenous dosing. Afterwards, clinical trials start only for those drug candidates that passed throughout all previous screening cascade and that presented good pharmacological properties during preclinical development. The clinical trials include four different phases (phases I-IV) in which the information obtained from early studies is used to support the design of future studies. Thus, it is important to recognise that safety, pharmacokinetics, efficacy and tolerability are considered during all phases of clinical trials, but one of these aspects may be predominant at a determined moment of clinical development or post-marketing pharmacovigillance [55].

Among all these techniques and methods, it is important to choice the right ones bearing in mind the intended application for that they are being used. In fact, at the early stages of drug discovery it is not essential to obtain complete pharmacokinetic profiles but rather a rank-order of compounds to further elimination those with poor pharmacokinetic properties. On the other hand, at the development stage more complete pharmacokinetics and longer-term safety studies must be done in animals and humans before marketing approval. For that reason, approaches such as N-cassette dosing or pooling samples, while acceptable and helpful in drug discovery, cannot be applied in preclinical and clinical development, increasing also at this level the bioanalytical requirements in order to obtain accurate and high-quality data.

BIOANALYSIS IN DRUG DISCOVERY AND DEVELOPMENT

Bioanalysis involves frequently the quantitative determination of small molecules such as drugs or their metabolites in biological matrices [56,57]. Taking into account all those aspects and procedures previously described in the current process of DDD, the great interest of bioanalysis to support drug development and drug discovery phases is undoubted. Indeed, in vitro screening assays usually performed at the drug discovery stage to predict ADME properties of drug candidates require their quantitative determination, as well as in later pharmacokinetic studies. Therefore, the development of sound bioanalytical methods is of paramount importance during the process of DDD. Nowadays, it is widely recognised that the bioanalysis is essential for pharmacokinetic/pharmacodynamic characterization of a drug. However, the bioanalytical methods developed to support the drug discovery stage may not be suitable for drug development and progressive modifications may be required. In fact, the degree of development of bioanalytical assays used in the discovery and development of NCEs tends to increase as the lead candidate progresses to more advanced stages. Actually, during the drug discovery phase, straightforward, rapid, efficient, and high-throughput analytical protocols are needed to tackle the elevated number of samples and to enhance the rate of drug candidates entering into drug development. However, at the same time that drug candidates pass to the next phase of DDD the quality of bioanalytical method should be improved and demonstrated by appropriate validation assays, which confirm its reliability, robustness, and accuracy. Effectively, the throughput of a method may decrease during the drug development phase but the quality of data generated must be higher.

All changes in the process of DDD have inevitably induced a continuous progress in the bioanalytical field to respond to the new requirements. Hence, attending the low concentrations and reduced sample volume generally available from HTS assays, the pharmaceutical industries were forced to adopt new analytical tools, making considerable progresses especially in sample pre-treatment, compound chromatographic separation, and detection conditions. In most biological fluids and tissues their endogenous compounds may interfere with the NCEs to be analysed. Therefore, a sample pre-treatment procedure should be developed and optimized to assure efficient sample clean-up and selective extraction of the analytes of interest without excessive losses. Protein precipitation (PPT) is the simplest means of sample pre-treatment as it involves only the addition of a precipitating solvent (methanol, acetonitrile or perchloric acid solution) with subsequent homogenization and centrifugation.

Although extensively used since it is fast and easily automated, many times PPT procedures are not as efficient as intended given that many matrix components remain in suspension, leading to high variations among samples [58,59]. Liquid-liquid extraction (LLE) may be used to overcome this limitation, in particular for lipophilic molecules because they are transferred from aqueous to an apolar organic phase [60-62]. This extraction procedure provides an excellent sample clean-up; however its use is difficult in an automated high-throughput format. Presently, the solid-phase extraction (SPE) procedure is the most widespread used, probably because it is versatile and easily automated [63-65]. Apart from full automation, SPE has demonstrated higher reproducibility, sensitivity and throughput when compared to LLE or PPT [64]. Other extraction procedures have been employed, however not so frequently: solid-phase microextraction [66, 67], supercritical fluid extraction [68], and methods using molecularly imprinted polymers [69].

Over the years, bioanalysis remains dominated by liquid chromatography (LC). LC has shown to be a powerful analytical tool for separation and quantification and it is of the utmost importance in quality control of pharmaceuticals and in bioanalysis of drugs and their metabolites, remaining, therefore, as the first choice for the determination of NCEs during DDD process. The reverse phase-liquid chromatography (RP–LC) is the most widely employed chromatographic technique due to its application to small molecules, which are separated by their differential relative affinity for the hydrophobic stationary phase in relation to mobile phase. LC has been in continuous development in order to reduce the run time and increase the throughput analysis without sacrificing the chromatographic separation and selectivity. Therefore, a large variety of chromatographic stationary phases have been developed both in the field of achiral and chiral chromatography. For example, in addition to the commonly used C_{18} silica columns, other packing materials can be employed in RP-LC: shorter alkyl chains such as C_{12} [70] or C_8 [71] and longer alkyl chains such as C_{30} [72]. Nevertheless, for NCEs with low octanol-water partition coefficient, it is not ease to establish a reliable RP-LC method because they will present short or no retention time. It is well known that drug metabolites normally possess higher polarity than their parent compounds and, consequently, it is difficult to achieve appropriate retention times. Thus, porous graphitic carbon chromatography [73,74], hydrophilic interaction chromatography [75], and ion pair chromatography [76] emerged as techniques which provide more efficient retention of polar compounds and have greater potential to separate polar metabolites [77]. Other approaches related to chromatographic techniques may be taken into account such as shorter length columns and smaller particle diameter. If the particle size is below 2 μm there is a substantial increase of the column backpressure, which may be technically solved by the development of instrumentation and columns surviving such high pressures. This technique is called ultra-performance liquid chromatography and it has been widely used in recent years [78-81]. Additionally, monolithic columns have been created and applied in LC [82,83]. They have a macroporous structure which allows very high flow rates (upper than 9 mL/min) without considerably sacrificing efficacy and method robustness. This is particularly interesting for high-throughput analysis, given that several compounds can be analysed in the same run which has a shorter time due to the high flow rate applied. Over the last years, the progress achieved in chiral separation and analysis was also wonderful and the presence of chiral centres in NCEs is not a major restriction for pharmaceutical industry at the discovery stage. On the other hand, as the chiral compound is advancing to the development stage, it is essential to know the pharmacological properties of each enantiomer because they may have different pharmacodynamics, pharmacokinetics and toxicity. In fact, the development of chiral separation methods is very important to investigate the occurrence of enantioselectivity in pharmacokinetics. The enantiomeric resolution and the quantification of enantiomers of chiral drugs may be performed by LC either indirectly using chiral derivatization reagents or directly using chiral selectors as mobile phase additives or bonded on the surface of a solid support (chiral stationary phase) [84-86]. LC has also coupled with success to several detectors [photodiode-array, ultra-violet, fluorescence, polarimetry and mass spectrometry (MS)] offering a more comprehensive analytical information. The ideal detector should yield a linear relationship between response and analyte concentration as well as great selectivity and sensitivity and structural information of drug candidates and their metabolites. Bioanalysis based on liquid chromatography-mass spectrometry (LC-MS) combines the efficient separation of chromatographic columns with the MS detector power. This technique received considerable attention because it allows high-throughput analysis without compromising the quality of data. Moreover, due to their inherent selectivity, LC-MS and LC-MS-MS systems normally require neither labour-intensive sample preparation nor long chromatographic run time. The extraordinary growth of LC-MS applications on quantitative bioanalysis is well documented [87,88], and several methodologies were successfully developed to achieve shorter run times for small molecules: direct injection [89], microcolumns [90], monolithic columns [91] and high-pressure columns [92,93]. Although today

MS is considered the most widespread tool for identification and detection of drugs and their metabolites, ultraviolet detection should not be underestimated. Indeed, when a molecule has a strong chromophoric group, it absorbs strongly in ultraviolet and, therefore, it may be easily detected and quantified. Liquid chromatography-ultraviolet detection (LC-UV) is often used in bioanalysis, especially to support in vitro bioavailability and metabolic stability screening [94-97] or in cases where sensitivity degree is not paramount or where LC-MS is not economically viable. Furthermore, ultraviolet detection may also be enough to differentiate positional isomers and enantiomers and it can be applied during the drug development stages [98,99].

CHANGES IN BIOANALYTICAL METHODS DURING DRUG DISCOVERY AND DEVELOPMENT

The field of bioanalysis covers a very broad range of assays. Of particular interest is the availability of reliable quantitative bioanalytical methods to support drug development. However, the need of appropriate quantitative bioanalytical methods takes place earlier at the level of drug discovery, but there are no clear guidelines to announce the minimal parameters to study in order to demonstrate which degree of method development is acceptable when it is used for analytical discovery purposes. Until now, in our knowledge, only Srinivas [100] discussed the changes that need to be made in a bioanalytical assay during the passage of a NCE from the early discovery stage as it becomes a preclinical candidate and/or clinical candidate and through the several phases of clinical development reaching the filling stage for a new drug. Taking the literature as well as our knowledge into account, the method validation requirements suggested during drug discovery and development are presented in Table 1. The development of sound and validated bioanalytical methods is of paramount value in whole DDD process. Nevertheless, this does not mean that the same method and validation parameters are considered at the different stages of DDD. Indeed, throughout the process of DDD the intended goals differ from one phase to another as so as the demands of validation, which are more rigorous during the preclinical and clinical development. Accordingly, in spite of all bioanalytical methods should afford enough confidence in the results that will be generated by itself, the validation parameters and their acceptance criteria should be firstly titrated in order to reach an optimal balance between speed and required quality of data. Otherwise, the drug discovery strategy will not be economically viable.

The major issues to deal with are related to the selectivity of the method and drug stability in solutions and biological matrices and after freeze/thaw cycles. Furthermore, accuracy and precision should be also evaluated but the acceptance criteria are larger at drug discovery phase than those applied during drug development. At the drug discovery stage a calibration curve must also be constructed for each analyte in an adequate concentration range to achieve reliable results, but the number of standards considered is smaller than those in drug development [100-102]. As the molecule advances into preclinical development, validation must be formalized respecting the international guidelines. Effectively, the bioanalytical data achieved in preclinical development should be unambiguous and trusty since they may influence the design of clinical trials and, sometimes, they may integrate the regulatory submission process. In general, it is incorporated an internal standard (IS) to compensate errors occurring at some point of the sample extraction procedure and/or chromatographic analysis. Regardless the method of extraction adopted, the IS must belong to the same chemical scaffold, presenting only minor structural differences when compared to the analyte. Validation experiments should be performed maintaining the concentration of the IS constant and the chromatographic response is given by analyte/IS peak area or height ratio [85,86]. Although the method validation should be very methodical during the preclinical phase, it needs to be even more rigorous to support clinical development [103-105]. In this case the bioanalytical method must be more robust and the selectivity and stability must be evaluated again. It is common that patients are concomitantly ingesting other dugs and, therefore, the assay must be flexible enough to accommodate minor alterations in chromatographic conditions to circumvent the interfering peaks if necessary [106]. Thereby, potential drug-drug or metabolite-drug interactions should be investigated. For this reason, co-prescribed drugs and metabolites must be included during the method development and validation [107,108]. Finally, besides plasma and serum, the method should be also easily adaptable to other biological matrices, such as urine [109].

As stated previously, the quality data required scales along the drug discovery and development. In the next section, it will be discussed the general recommendations for LC bioanalytical method validation employed for the quantitative determination of drugs and/or metabolites in biological matrices, to support bioavailability, bioequivalence, and pharmacokinetic studies in man and animals.

Table 1. Bioanalysis and method validation requirements suggested during drug discovery and development.

Stage	Characteristics	Objectives	Validation
Drug Discovery	• Very fast • High-throughput • Straightforward • Moderate quality data	• Absorption screening (n-octanol/water, PAMPA, Caco-2 cells) • Metabolic stability screening (microsomes, hepatocytes) • In vitro cytochrome P450 inhibitory screening (microsomes, human recombinant enzymes) • Plasma protein binding (ultrafiltration, equilibrium dialysis) • In vivo pharmacokinetic screening (cassette dosing, pooling samples)	• Early study of detection and chromatographic conditions • Stability assays of stock solution • Former sample extraction procedure • Selectivity (no sample endogenous interferences, no interference of co-administered leads in cassette dosing and/or cassette analysis) • Calibration curve (low number of standards, n=3-5) • Preliminary quantification range • Reasonable precision and accuracy
Preclinical Drug Development	• Fast • Moderate-throughput • High quality data	• Pharmacokinetic characterization in rodents, dogs, monkeys (single and multiple dose studies, absolute bioavailability) • Route-dependent pharmacokinetic disposition (intravenous/oral) • Toxicokinetics characterization of parent drug and its metabolites in toxicology species (dose/exposure levels, estimating the first dose to human exposure) • In vivo drug-drug interaction potential	• Sample extraction procedure optimization • Selectivity (no sample endogenous interferences, no interference of parent drug/metabolites) • Internal standard selection • Defined range of standard calibration curve • LLQ and ULQ establishment • Intra and inter-day precision and accuracy assessment (3-5 validation runs) • Effect of dilution on precision and accuracy • Extraction recovery assessment (parent drug/metabolites and internal standard) • Stability experiments to cover sample handling and storage conditions

Table 1. (Continued)

Stage	Characteristics	Objectives	Validation
Clinical Drug Development	• As fast as possible • Low-throughput • Very high quality data	• Pharmacokinetics in healthy humans, target patient population and special populations (paediatric, geriatric, hepatic and renal impairment) • Food effect and relative bioavailability (solution/suspension vs solid dosage form) • Drug-drug interactions with agents commonly co-prescribed or with narrow safety window	• Finalization of extraction, chromatographic and detection conditions (based on anticipated concentrations following the lowest dose in humans) • Full validation is required: selectivity, linearity, LOQ, LOD, accuracy, precision, recovery, parent drug/metabolites stability experiments • Verification of robustness/ruggedness of the assay

BIOANALYTICAL METHOD VALIDATION: FROM A REGULATORY PERSPECTIVE

Increasing interdependence among the countries has emerged as a driving force for the standardization of many aspects of common interest. Consequently, also in the bioanalysis field, from the beginning of nineties, in order to assure a common (minimum) level of quality data and their international acceptability, several recommendations have been successively published related to bioanalytical method validation. Indeed, the results generated by analytical methods not validated are devoid of significance and will not be recognised by official authorities. Today, it is widely acknowledged that the employment of well-characterised and fully validated analytical methods is the key determinant to yield reliable and reproducible results which may be reasonable interpreted [110].

Method validation is the systematic procedure required to demonstrate that an analytical method is acceptable for its intended application. At this point, it must be stated out that the performance of an analytical method largely depends on its development degree and not so much on its validation. In fact, it is impossible to separate analytical method validation and development, since they represent an iterative process where the validation studies determine if the status of method development is satisfactory or further changes and revalidation are needed. Therefore, in general, a well-developed method should be easy to validate.

Nowadays, the validation of bioanalytical methods to quantify small molecules in biological samples is based on a conference report published in 2000 by Shah et al. [111] on *Bioanalytical Method Validation – A Revisit with a Decade of Progress* which constitutes an update to the first conference on *Analytical Methods Validation: Bioavailability, Bioequivalence and Pharmacokinetic Studies* [112]. Following that, in May 2001 the Food and Drug Administration (FDA) issued the present Guidance for Industry on bioanalytical method validation [113], which was developed based on the deliberations of the two conferences referred to previously. These documents close the current guiding principles worldwide followed by bioanalytical laboratories for validation of bioanalytical methods used to support bioavailability, bioequivalence, and pharmacokinetic studies, and include also a reference for the regulatory agencies. As stated in these documents, a full validation of a bioanalytical method is necessary when developing and implementing the method for the first time, when it is required for a new drug entity or whenever metabolites are added to an assay formerly used for quantification.In other circumstances only a partial validation may be needed. However, herein, it will be done an overview focusing the validation parameters and their acceptance criteria as well as the validation procedures required for the development of a new bioanalytical method. For quantitative bioanalytical procedures there is a general agreement that the subsequent validation parameters should be evaluated: selectivity, calibration model (linearity), limit of quantification (LOQ), precision, accuracy, recovery, and stability of the analytes in the biological matrix to be assessed.

Selectivity is referred to as the ability of a bioanalytical method to unequivocally differentiate and quantify the analytes of interest in the presence of other sample components. Endogenous matrix components, metabolites, and decomposition products or even co-prescribed drugs include the most common interferences. Thus, it is necessary to establish that the chromatographic signal produced is only due to the analyte and not as a result of the co-elution of other compounds. Many times, the terms selectivity and specificity appear interchangeably in the literature, but the use of the term selectivity has been recommended in detrimental of the term specificity [114]. In fact, few, if any, methods are specific (100% selective). The most part of chromatographic methods produce responses not only for the analyte, but also for other components, being the use of the term selectivity more correct in this context. A bioanalytical method is considered to be selective if the lack of response in the blank biological matrix at the retention time of the analytes is demonstrated. As discussed in the FDA Guidance [113] blank samples of appropriate biological matrix should be analysed from at least six independent sources. Each sample should be tested and the selectivity should be ensured at the LOQ. The Figure 2 was obtained from a study conducted recently by our group where it is exemplified the lack of response in blank human plasma at the retention time of the analytes of interest: internal standard, R-licarbazepine, S-licarbazepine, oxcarbazepine and eslicarbazepine acetate [98].

The quality of the calibration curve is an aspect of the greatest importance in bioanalytical method validation. In fact, as Almeida et al. [115] stated, the quality of bioanalytical data is highly dependent on the quality of the calibration curve used to extrapolate the analyte concentrations in unknown samples. The calibration curve is the relationship between instrumental response and the concentrations of the analyte, and it must be generated for each analyte, using a sufficient number of calibration standards. Taking into account the FDA Guidance [113], the calibration curve should be prepared in the same biological matrix as the samples in the intended study and it should be constructed with at least six calibration standards covering the expected concentration range, including the LOQ (lowest concentration of a sample that may still be quantified with satisfactory precision and accuracy). The calibration model most commonly adopted is the unweighted linear regression, which is appropriate if there is homogeneity of variances across the calibration range. However, this condition is expected only for calibration ranges spanning not more than one order of magnitude and it is frequently not met for analytical data obtained from pharmacokinetic studies. Therefore, when heteroscedasticity condition is demonstrated a simple and effective way to overcome the greater influence of the highest concentrations on the fitted regression line is to use a weighted least squares linear regression model. The steps to be taken in the selection of best linear calibration model were exhaustively described by Almeida et al. [115] and this was illustrated by a practical example using a data set obtained during the validation process of a LC-UV method for the determination of lamotrigine in plasma. More recently, we also used a weighted least squares linear regression model in the validation of enantioselective LC-UV methods to quantify eslicarbazepine acetate and its metabolites in human plasma and in four biological matrices of mouse [98,116]. Usually, the heteroscedasticity is adequately compensated by the inverse of the concentration ($1/x$) or the inverse of the squared concentration ($1/x^2$), but statistical aspects should be considered in the selection of the most appropriate weighting factor. Nevertheless, in bioanalytical method validation the choice of weighting should be justified, since the guidelines recommend the use of the simplest model that adequately describes the concentration-response relationship. At last, after selection of the adequate calibration model the linearity range must be defined. The recommended validation procedures and the suggested acceptance criteria for the calibration curve and LOQ are summarized in the Table 2.

Precision and accuracy are key validation parameters of any analytical method and they reflect the status of the method development. In fact, to obtain analytical data with good precision and accuracy it is necessary to assure an appropriate procedure for sample collection and extraction, an adequate IS and chromatographic conditions that minimize interferences, as well an enough signal-noise to permit reproducible peak integration. The precision of an analytical method expresses the closeness degree among individual measures of an analyte when the same procedure is applied repeatedly to multiple aliquots of a homogenous matrix, and it may be considered at three levels: intra-day precision or repeatability, inter-day precision or intermediate precision and reproducibility. Intra-day precision expresses the precision under the same experimental conditions over a short interval of time. Inter-day precision expresses variations in the same laboratory: different days, different analysts, etc. Reproducibility expresses the precision between laboratories and only has been studied if the method is supposed to be employed in different laboratories. On the other hand, the accuracy describes the closeness between the mean experimental data obtained by the method and the corresponding nominal concentration. Hence, accuracy is determined analysing replicate samples of known amounts of the analyte. From a practical point of view, the precision and accuracy of an analytical method can be estimated from the analysis of quality control (QC) samples, which represent the future real samples. QC samples should be obtained independently of the batch of calibration standards and, if possible, they should not be prepared by the same person performing the validation study and should be representative of the whole calibration range. In this context, one factor that can influence the precision and accuracy is the recovery of the analytes and IS which must to be investigated. In current guidelines on bioanalytical method validation none numerical value is specified in relation to the percentage of recovery, but they state that the recovery of the analytes and IS should be consistent, precise and reproducible. The recommended validation experiments to examine the precision, accuracy and recovery of an analytical method and their acceptance criteria are also presented in the Table 2.

Stability of the analytes throughout the analytical procedure is also a prerequisite for their reliable quantitative determination. Thus, a full validation of a bioanalytical method must include stability experiments during sample collection and handling as well as under storage conditions prior to analysis. In general, the analyte stability experiments should involve short-term stability studies (bench top, room temperature), long-term stability studies (in the frozen biological matrix at the intended storage temperature), freeze/thaw cycles stability studies (reanalysis), post-preparative stability (residence time in the autosampler) and stock solution stability studies. Drug stability in a biological sample will be function of the storage conditions, the chemical properties of the drug, the matrix and the nature of container system. Therefore, the stability data of an analyte in a particular condition should not be extrapolated to other conditions. Experimentally, the stability of the analytes is assessed comparing the chromatographic response obtained from samples analysed before (reference samples) and after being exposed to the conditions for stability assessment (stability samples). Concerning this topic more useful information is shown in the Table 2.

Table 2. Bioanalytical method validation: parameters and their acceptance criteria as well as recommended assays.

Parameter	Acceptance Criteria	Assays
Selectivity	No significant interference including at the LOQ	At least 6 blank samples from independent sources
Calibration curve	Analyte response at the LOQ at least 5 times the blank response; Coefficient of variation ≤15% (≤20% at LOQ) and mean concentration within ±15% of the nominal value (±20% at LOQ); At least 4/6 non-zero standards should meet the above criteria	Blank sample, zero sample (with IS), 6-8 spiked calibration samples covering the entire calibration range; Apply the simplest calibration model
LOQ	Analyte response at least 5 times the blank response; Coefficient of variation ≤20% and mean concentration within ±20% of the nominal value	Analyse at least 5 samples independent of standards
Precision	Coefficient of variation ≤15% (≤20% at LOQ)	Minimum of 3 concentrations representing the entire calibration range (near LOQ, mid-range and near upper end of range) and at least 5 replicates per concentration
Accuracy	Mean concentration within ±15% of the nominal value (±20% at LOQ)	Minimum of 3 concentrations representing the entire calibration range (near LOQ, mid-range and near upper end of range) and at least 5 replicates per concentration
Recovery	Analyte and IS recovery should be consistent, precise and reproducible	Compare extracted samples to unextracted standards at 3 concentrations representing the entire calibration range (near LOQ, mid-range and near upper end of range)
Short-term stability	No specific criteria	Assess at 2 concentrations represent-ting the entire calibration range (low and high) and at least 3 replicates per concentration; Thaw and store at room temperature for 4-24 h
Long-term stability	No specific criteria	Assess at 2 concentrations representing the entire calibration range (low and high) and at least 3 replicates per concentration; From date of first sample collection to last sample analysis.

Table 2. (Continued)

Parameter	Acceptance Criteria	Assays
Freeze/thaw stability	No specific criteria	Assess at 2 concentrations representing the entire calibration range (low and high) and at least 3 replicates per concentration; Store at intended temperature for 24 h, thaw unassisted at room temperature, freezing for 12-24 h and repeat twice more.
Post-preparative stability	No specific criteria	Assess in the autosampler conditions for largest batch size
Stock solution stability	No specific criteria	Assess at room temperature for at least 6 h; Document stability for the intended time if stored refrigerated or frozen

Table 3. Run acceptance criteria for the routine analysis.

Parameter	Acceptance Criteria	Assays
Calibration curve	75% or a minimum of 6 calibration samples must be within ±15% of the nominal value (±20% at LOQ). Values outside these limits may be discarded if the model is not affected	Minimum of 6 calibration samples covering the entire concentration range, excluding blanks. The same calibration model determined during the method validation should be used.
QC samples	Mean concentration must be within ±15% of the nominal value for at least 67% (4/6); 2 QC samples may fail the above criteria, but not at the same concentration level	Assess in duplicate at 3 concentrations representing the entire calibration range (near LOQ, mid-range and near upper end of range); QC samples should comprise at least 5% of unknown samples, but no less than 6.

Besides the full validation of a bioanalytical method, when it is used routinely a calibration curve should be constructed for each analyte in each analytical run. In addition, in routine drug analysis the precision and accuracy should be monitored to ensure that the method performance remains satisfactory. For that, QC samples prepared separately should be analyzed with processed test samples and the results of the QC samples provide the basis for run acceptation or rejection. In Table 3 are specified the run acceptance criteria for the routine analysis.

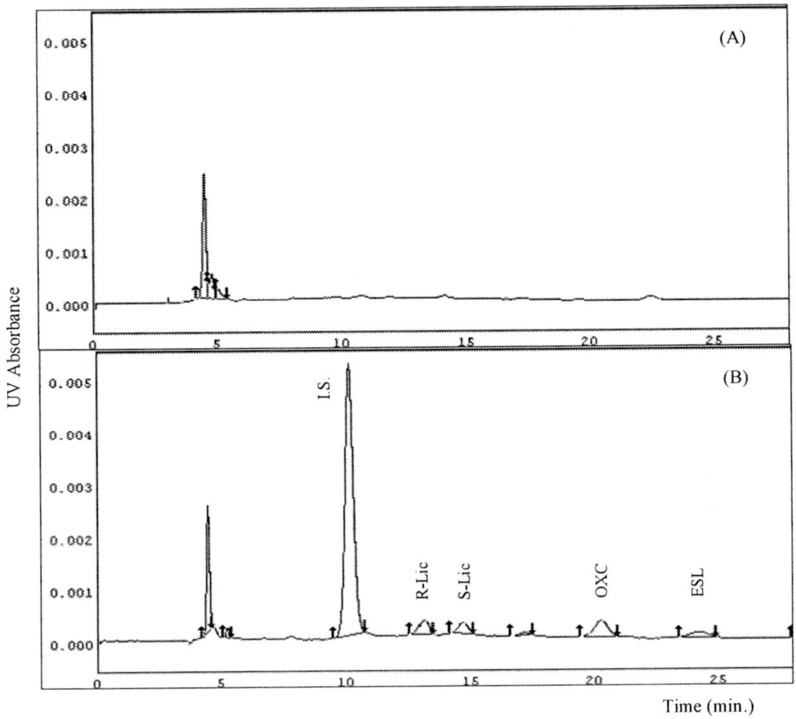

Figure 2. Typical chromatograms of extracted human plasma: (A) blank human plasma; (B) human plasma spiked with internal standard (I.S.) (conc.= 8 μg/mL) and drugs of interest at concentrations equivalent to the LOQ [R-licarbazepine (R-Lic), S-licarbazepine (S-Lic), oxcarbazepine (OXC) and eslicarbazepine acetate (ESL) (conc.= 0.4 μg/mL)].

CONCLUSION

The development and validation of sound bioanalytical methods to support the DDD process is determinant to achieve high-quality data, contributing for a reliable interpretation of pharmacological phenomena. Today, it is well-recognised that a bioanalytical method undergoes successively changes and improvements over the process of DDD and even after marketing approval of a new drug. Indeed, many times, the status of development of a bioanalytical method follows the technological advances, which will become the bioanalytical field continuouslly more regulated.

ABBREVIATIONS

ADME	Absorption, distribution, metabolism and excretion
DDD	Drug discovery and development
FDA	Food and Drug Administration
HTS	High-throughput screening
IS	Internal standard
LC	Liquid chromatography
LC-MS	Liquid chromatography-mass spectrometry
LC-MS-MS	Liquid chromatography coupled with tandem mass spectrometry
LC-UV	Liquid chromatography-ultraviolet detection
LLE	Liquid-liquid extraction
LOQ	Limit of quantification
MS	Mass spectrometry
NCEs	New chemical entities
PAMPA	Parallel artificial membrane permeability assay
PPT	Protein precipitation
QC	Quality Control
RP-LC	Reverse phase-liquid chromatography
SPE	Solid-phase extraction
UV	Ultraviolet

REFERENCES

[1] Gao, L; Cheng, X; Zhang, J; Burns, DJ. *Rapid Commun Mass Spectrom.*, 2007, 21, 3497-3504.

[2] Inglese, J; Johnson, RL; Simeonov, A; Xia, M; Zheng, W; Austin, CP; Auld, DS. *Nat Chem Biol.*, 2007, 3, 466-479.

[3] Snowden, M; Green, DV. *Curr Opin Drug Discov Devel.*, 2008, 11, 553-558.

[4] Srinivas, NR. *Biomed Chromatogr.*, 2006, 20, 383-414.

[5] Curatolo, W. *Pharm Sci Tech Today*,1998, 1, 387-393.

[6] Venkatesh, S; Lipper, RA. *J Pharm Sci.*, 2000, 89, 145-154.

[7] Caldwell, GW; Ritchie, DM; Masucci, JA; Hageman, W; Yan, Z. *Curr Top Med Chem.*, 2001, 353-366.

[8] van Waterbeemd, H; Gifford, E. *Nat Rev Drug Discov.*, 2003, 2, 192-204.

[9] Ismail, K; Landis, J. *Nat Rev Drug Discov.*, 2004, 3, 711-716.

[10] Fotouhi, N; Gillespie, P; Goodnow, RA; So, SS; Han, Y; Babiss, LE. *Comb Chem High Throughput Screen.*, 2006, 9, 95-102.

[11] Fortuna, A; Alves, G; Falcão, A. *Curr Top Pharmacol.*, 2007, 11, 63-86.

[12] Yengi, LG; Leung, L; Kao, J. *Pharm Res.*, 2007, 24, 842-858.

[13] Kola, I; Landis, J. *Nat Rev Drug Discov.*, 2003, 3, 711-715.

[14] Kansy, M; Senner, F; Gubernator, K. *J Med Chem.*, 1998, 41, 1007-1010.

[15] Avdeef, A; Strafford, M; Block, E; Balogh, MP; Chambliss, W; Khan, I. *Eur J Pharm Sci.*, 2001, 14, 271-280.

[16] Wohnsland, F; Faller, B. *J Med Chem.*, 2001, 44, 923-930.

[17] Sugano, K; Hamada, H; Machida, M; Ushio, H; Saitoh, K; Terada, K. *Int J Pharm.*, 2001, 228, 181-188.

[18] Sugano, K; Takata, N; Machida, M; Saitoh, K; Terada, K. *Int J Pharm.*, 2002, 241, 241-251.

[19] Zhu, C; Jiang, L; Chen, TM; Hwang, KK. *Eur J Med Chem.*, 2002, 37, 399-407.

[20] Sugano, K; Nabuchi, Y; Machida, M; Aso Y. *Int J Pharm.*, 2003, 257, 245-251.

[21] Di, L; Kerns, EH; Fan, K; McConnell, OJ; Carter, GT. *Eur J Med Chem.*, 2003, 38, 223-232.

[22] Malakoutikhah, M; Teixidó, M; Giralt, E. *J Med Chem.*, 2008, 51, 4881-4889.

[23] Ottaviani, G; Martel, S; Carrupt, PA. *J Med Chem.*, 2006, 49, 3948-3954.

[24] Verma, RP; Hansch, C; Selassie, CD. *J Comput Aided Mol Des.*, 2007, 21, 3-22.

[25] Sandhya, KV; Devi, GS; Mathe, ST. *Mol Pharm.*, 2008, 5, 92-97.

[26] Serra, H; Mendes, T; Bronze, M; Simplício, A. *Bioorg Med Chem.*, 2008, 16, 4009-4018.

[27] Markowska, M; Oberle, R; Juzwin, S; Hsu, CP; Gryszkiewicz, M; Streeter, AJ. *J Pharmacol Toxicol Methods*, 2001, 46, 51-55.

[28] Palmgrén, JJ; Mönkkönen, J; Jukkola, E; Niva, S; Auriola, S. *Eur J Pharm Biopharm.*, 2004, 57, 319-328.

[29] Thomas, S; Brightman, F; Gill, H; Lee, S; Pufong, B. *J Pharm Sci.*, 2008, 97, 4557-4574.

[30] Balimane, PV; Han, YH; Chong, S. *AAPS Journal*, 2006, 8, E1-E13.

[31] Cheng, D; Iqbal, J; Devenny, J; Chu, CH; Chen, L; Dong, J; Seethala, R; Keim, WJ; Azzara, AV; Lawrence, RM; Pelleymounter, MA; Hussain, MM. *J Biol Chem.*, 2008, 283, 29802-29811.

[32] Iwan, M; Jarmołowska, B; Bielikowicz, K; Kostyra, E; Kostyra, H; Kaczmarski, M. *Peptides*, 2008, 29, 1042-1047.

[33] Siissalo, S; Hannukainen, J; Kolehmainen, J; Hirvonen, J; Kaukonen, AM. *Eur J Pharm Biopharm*, 2009, 71, 332-338.

[34] Tan, W; Chen, H; Zhao, J; Hu, J; Li, Y. *J Pharm Pharm Sci.*, 2008, 11, 97-105.

[35] Guo, B; Li, C; Wang, G; Chen, L. *Rapid Commun Mass Spectrom.*, 2006, 20, 39-47.

[36] Motoya, T; Thevanayagam, LN; Blashke, TF; Stone, JA; Jayewardene, AL; Aweeka, FT. *HIV Med.*, 2006, 7, 122-128.

[37] Clarke, HJ; Gregoire, F; Ma, F; Martin, R; Zhao, S; Lavan, BE. *PPAR Res.*, 2008, 465715.

[38] Zhu, CJ; Zhang, JT. *Chirality*, 2009, 21, 402-406.

[39] Wilson, ID; Nicholson, JK. *Xenobiotica*, 2003, 33, 887-901.

[40] Hollenberg, PF; Kent, UM; Bumpus, NN. *Chem Res Toxicol.*, 2008, 21, 189-205.

[41] Brown, HS; Galetin, A; Hallifax, D; Houston, JB. *Clin Pharmacokinet.*, 2006, 45, 1035-1050.

[42] Lu, C; Miwa, GT; Prakash, SR; Gan, LS; Balani, SK. *Drug Metab Dispos.*, 2007, 35, 79-85.

[43] Ohno, Y; Hisaka, A; Suzuki, H. *Clin Pharmacokinet.*, 2007, 46, 681-696.

[44] Han, HK; Sadagopan, N; Reichard, GA; Yapa, U; Zhu, T; Hubbel, A; Johnson, K; Brodfuehrer, J. *J Pharm Sci.*, 2006, 95, 1684-1692.

[45] Chris,t, DD. Cassette Dosing Pharmacokinetics: Valuable Tool or Flawed Science? *Drug Metab Dispos.*, 2001, 29, 935.

[46] Liu, B; Chang, J; Gordon, WP; Isbell, J; Zhou, Y; Tuntland, T. *Drug Discov Today*, 2008, 13, 360-367.

[47] Manitpisitkul, P; White, RE. *Drug Discov Today*, 2004, 9, 652-658.

[48] Raynaud, FI; Fischer, PM; Nutley, BP; Goddard, PM; David, P; Lane, DP; Workman, P. *Mol Cancer Ther.*, 2004, 3, 353-362.

[49] Mei, H; Korfmacher, W; Morrison, R. *AAPS Journal*, 2006, 8, E493-E500.

[50] Smith, NF; Raynaud, FI; Workman, P. *Mol Cancer Ther.*, 2007, 6, 428-440.

[51] He, K; Qian, M; Wong, H; Bai, SA; He, B; Brogdon, B; Grace, JE; Xin, B; Wu, J; Ren, SX; Zeng, H; Deng, Y; Graden, DM; Olah, TV; Unger, SE; Luettgen, JM; Knabb, RM; Pinto, DJ; Lam, PY; Duan, J; Wexler, RR; Decicco, CP; Christ, DD; Grossman, SJ. *J Pharm Sci.*, 2008, 97, 2568-2580.
[52] Ward, KW; Proksch, JW; Levy, MA; Smith, BR. *Drug Metab Dispos.*, 2001, 29, 82-88.
[53] Cheung, BW; Cartier, LL; Russlie, HQ; Sawchuk, RJ. *Fundam Clin Pharmacol.*, 2005, 19, 347-354.
[54] Gombar, VK; Silver, IS; Zhao, Z. *Curr Top Med Chem.*, 2003, 3, 1205-1225.
[55] Dingemanse, J; Appel-Dingemanse, S. *Clin Pharmacokinet.*, 2007, 46, 713-737.
[56] Deng, Y; Wu, JT; Lloyd, TL; Chi, CL; Olah, TV; Unger, SE. *Rapid Commun Mass Spectrom.*, 2002, 16, 1116-1123.
[57] Schellen, A; Ooms, B; Lagemaat, D; Vreeken, R; Dongen, WD. *J Chromatogr B*, 2003, 788, 251-259.
[58] Gangl, ET; Annan, M; Spooner, N; Vouros, P. *Anal Chem.*, 2001, 73, 5635-5644.
[59] Wille, SM; Lambert, WE. *Anal Bioanal Chem.*, 2007, 388, 1381-1391.
[60] Ho, TS; Pedersen-Bjergaard, S; Rasmussen, KE. *J Chromatogr A*, 2002, 963, 3-17.
[61] Pedersen-Bjergaard, S; Rasmussen, KE. *J Chromatogr B*, 2005, 817, 3-12.
[62] Pedersen-Bjergaard, S; Rasmussen, KE. *J Chromatogr A*, 2008, 1184, 132-142.
[63] Byrd, GD; Ogden, MW. *J Mass Spectrom.*, 2003, 38, 98-107.
[64] Shinozuka, T; Terada, M; Tanaka, E. *Forensic Sci Int.*, 2006, 162, 108-112.
[65] Castro, A; Concheiro, M; Quintela, O; Cruz, A; López-Rivadulla, M. *J Pharm Biomed Anal.*, 2008, 48, 183-193.
[66] Musteata, FM; Pawliszyn, J. *J Biochem Biophys Methods*, 2007, 70, 181-193.
[67] Cháfer-Pericás, C; Campíns-Falcó, P; Prieto-Blanco, MC. *Anal Chim Acta*, 2008, 610, 268-273.
[68] Radcliffe, C; Maguire, K; Lockwood, B. *J Biochem Biophys Methods*, 2000, 43, 261-272.
[69] Theodoridis, G; Konsta, G; Bagia, C. *J Chromatogr B*, 2004, 804, 43-51.
[70] Gréen, H; Vretenbrant, K; Norlander, B; Peterson, C. *Rapid Commun*

Mass Spectrom., 2006, 20, 2183-2189.
[71] Lampinen-Salomonsson, M; Beckman, E; Bondesson, U; Hedeland, M. *J Chromatogr B*, 2006, 833, 245-256.
[72] Moriwaki, H; Watanabe, A; Arakawa, R; Tsujimoto, Y; Shimizu, M; Noda, T; Warashina, M; Tanaka, M. *J Mass Spectrom.*, 2002, 37, 1152-1157.
[73] Xia, YQ; Jemal, M; Zheng, N; Shen, X. *Rapid Commun Mass Spectrom.*, 2006, 20, 1831-1837.
[74] Hsieh, Y; Duncan, CJ; Brisson, JM. *Rapid Commun Mass Spectrom.*, 2007, 21, 629-634.
[75] Naidong, W. *J Chromatogr B*, 2003, 796, 209-224.
[76] Hsieh, Y; Duncan, CJ. *Rapid Commun Mass Spectrom.*, 2007, 21, 573-578.
[77] Zhang, X; Rauch, A; Lee, H; Xiao, H; Rainer, G; Logothetis, NK. *Rapid Commun. Mass Spectrom.*, 2007, 21, 3621-3628.
[78] Dear, GJ; James, AD; Sarda, S. *Rapid Commun Mass Spectrom.*, 2006, 20, 1351-1360.
[79] Bae, SK; Kang, MJ; Yeo, CW; Kim, MJ; Shon, JH; Liu, KH; Shin, JG. *Biomed Chromatogr.*, 2008, 22, 939-946.
[80] Qin, F; Ma, Y; Wang, Y; Chen, L; Wang, D; Li, F. *J Pharm Biomed Anal.*, 2008, 46, 557-562.
[81] Zhang, SQ; Vinnakota, H; Jung, JC; Carvalho, P; Chittiboyina, AG; Avery, MA; Avery, BA. *Biomed Chromatogr.*, 2009, 23, 302-307.
[82] Tzanavaras, PD; Themelis, DG. *J Pharm Biomed Anal.*, 2007, 43, 1483-1487.
[83] Guddat, S; Thevis, M; Thomas, A; Schänzer, W. *Biomed Chromatogr.*, 2008, 22, 695-701.
[84] Gübitz, G; Schmid, MG. *Biopharm Drug Dispos.*, 2001, 22, 291-336.
[85] Srinivas, NR. *Biomed Chromatogr.*, 2004, 18, 207-233.
[86] Srinivas, NR. *Biomed Chromatogr.*, 2004, 18, 759-784.
[87] Xu, Y; Huang, J; Liu, F; Gao, S; Guo, Q. *J Chromatogr B*, 2007, 852, 101-107.
[88] Fang, AS; Miao, X; Tidswell, PW; Towle, MH; Goetzinger, WK; Kyranos, JN. *Mass Spectrom Rev.*, 2008, 27, 20-34.
[89] Hsieh, Y; Brisson, JM; Ng, K; Korfmacher, WA. *J Pharm Biomed Anal.*, 2002, 27, 285-293.
[90] Hsieh, Y; Brisson, JM; Wang, G; Ng, K; Korfmacher, WA. *J Pharm Biomed Anal.*, 2003, 33, 251-261.
[91] Zhang, K; Wu, S; Tang, X; Kaiser, NK; Bruce, JE. *J Chromatogr B*,

2007, 849, 223-230.

[92] Li, X; Fekete, A; Englmann, M; Götz, C; Rothballer, M; Frommberger, M; Buddrus, K; Fekete, J; Cai, C; Schröder, P; Hartmann, A; Chen, G; Schmitt-Kopplin, P. *J Chromatogr A*, 2006, 1134, 186-193.

[93] Li, KY; Zhou, YA; Ren, HY; Wang, F; Zhang, BK; Li, HD. *J Chromatogr B*, 2007, 850, 581-585.

[94] Shen, G; Hong, JL; Kong, AT. *J Chromatogr B*, 2007, 852, 56-61.

[95] Li, D; Wang, Q; Yuan, Z; Zhang, L; Xu, L; Cui, Y; Duan, K. *J Pharm Biomed Anal.*, 2008, 47, 429-434.

[96] Vorwerk, CK; Streit, F; Binder, L; Tuchen, S; Knop, C; Behrens-Baumann, W. *Graefes Arch Clin Exp Ophthalmol.*, 2008, 246, 1179-1183.

[97] Zhang, P; Li, F; Yang, XW. *Biomed Chromatogr.*, 2008, 22, 758-762.

[98] Alves, G; Figueiredo, I; Castel-Branco, M; Loureiro, A; Fortuna, A; Falcão, A; Caramona, M. *Biomed. Chromatogr.*, 2007, 21, 1127-1134.

[99] Trivedi, RK; Dubey, PK; Mullangi, R; Srinivas, NR. *J Chromatogr B*, 2007, 857, 224-230.

[100] Srinivas, NR. *Biomed Chromatogr.*, 2008, 22, 235-243.

[101] Briem, S; Martinsson, S; Bueters, T; Skoglund, E. *Rapid Commun Mass Spectrom.*, 2007, 21, 1965-1972.

[102] Wang, L; Sun, Y; Niu, W; Lu, T; Kan, J; Xu, F; Yuan, K; Qin, T; Liu, C; Li, C. *Rapid Commun Mass Spectrom.*, 2007, 21, 2573-2584.

[103] Frahnert, C; Rao, ML; Grasmäder, K. *J Chromatogr B*, 2003, 794, 35-47.

[104] Wittemer, SM; Veit, M. *J Chromatogr B*, 2003, 793, 367-375.

[105] Trivedi, RK; Kallem, RR; Mullangi, R; Srinivas, NR. *J Pharm Biomed Anal.*, 2005, 39, 661-669.

[106] Zhou, Z; Li, X; Xie, Z; Cheng, Z; Peng, W; Wang, F; Zhu, R; Li, H. *J Chromatogr B*, 2004, 802, 257-262.

[107] Afshar, M; Rouinin, MR; Amini, M. *J Chromatogr B*, 2004, 802, 317-322.

[108] Allqvist, A; Wennerholm, A; Svensson, JO; Mirghani, RA. *J Chromatogr B*, 2005, 814, 127-131.

[109] Jourdil, N; Bessard, J; Vicent, F; Eysseric, H; Bessard, G. *J Chromatogr B*, 2003, 788, 207-219.

[110] Hartmann, C; Smeyers-Verbeke, J; Massart, DL; McDowall, RD. *J Pharm Biomed Anal.*, 1998, 17, 193-218.

[111] Shah, VP; Midha, KK; Findlay, JW; Hill, HM; Hulse, JD; McGilveray, IJ; McKay, G; Miller, KJ. *Pharm Res.*, 2000, 17, 1151-1557.

[112] Shah, VP; Midha, KK; Dighe, S; McGilveray, IJ; Skelly, JP; Yacobi, A; Layloff, T; Viswanathan, CT; Cook, CE; McDowall, RD; Pittman, KA; Spector, S. *Pharm Res.*, 1992, 9, 588-592.
[113] Guidance for Industry, Bioanalytical method validation. *US Food and Drug Administration*, May 2001.
[114] Vessman, J; Stefan, RI; van Staden, JF; Danzer, K; Lindner, W; Burns, DT; Fajgelj, A; Muller, H. *Pure Appl Chem.*, 2001, 73, 1381-1386.
[115] Almeida, AM; Castel-Branco, MM; Falcão, AC. *J Chromatogr B*, 2002, 774, 215-222.
[116] Alves, G; Figueiredo, I; Castel-Branco, M; Loureiro, A; Falcão, A; Caramona, M. *Anal Chim Acta*, 2007, 596, 132-140.

In: Biomedical Chromatography
Editors: John T. Elwood, pp. 29-68
ISBN: 978-1-60741-291-5

Chapter 2

CHARACTERIZATION OF DRUG-PLASMATIC PROTEIN INTERACTIONS BY MICROSEPARATION TECHNIQUES

Maria Amparo Martínez-Gómez, Laura Escuder-Gilabert, Rosa M. Villanueva-Camañas, Salvador Sagrado and M. José Medina-Hernández*

Departamento de Química Analítica. Universidad de Valencia.
C/ Vicente A. Estellés s/n, E-46100, Burjassot Valencia, Spain

ABSTRACT

The study of the interaction of drugs with plasmatic proteins is essential for the understanding of the pharmacological properties of drugs. At the pharmacokinetic level, bound drug to proteins is a reserve of drug in the organism since it cannot be distributed, metabolized and excreted; so, only free drug is active since it can cross barriers and move towards tissues. The interaction of drugs with proteins can be of different types: hydrophobic, electronic or steric. In this chapter, the methodologies employed for characterizing these interactions are revised and compared. The interactions of β-blockers, phenothiazines, antihistamines, flavonoids, benzodiazepines and barbiturates with human serum albumin (HSA), α1-glycoprotein acid (AGP) and lipoproteins (LIPO) and/or globulins (GLOB) are evaluated by capillary electro-

* Corresponding author: e-mail: maria.j.medina@uv.es. Tel + 34963544899, Fax +34963544953.

phoresis and summarized. The use of chemometrics to characterize the nature of interactions is also reviewed.

1. INTRODUCTION

Human plasma contains more than 60 proteins, being albumin (HSA), α_1-acid glycoprotein (AGP), lipoproteins and globulins the most important from the point of view of the binding to drugs.

Albumin is the major plasmatic protein in the circulatory system since it represents 60% of the protein content of the plasma, with a concentration of 35–45 g/L (550–600 μM). It has a molecular weight of 66500 g/mol and an isoelectric point of 4.8 [1,2]. The primary structure of HSA consists of a simple polypeptidic chain of 585 amino acids that contains 17 disulphur bridges and a cisteine residue [3]. The secondary structure of HSA is spiral and is formed by the domains I, II and III that have different properties but its tertiary structure is very similar. Every domain is divided into the sub-domains A and B, which consist of 6 and 4 α-helices, respectively.

HSA molecule has at least six selective binding sites and a great number of low affinity binding sites [4]. Each drug preferably binds to a binding site although it can also bind to another site with minor affinity [5]. Several studies suggest that the binding sites are formed due to conformational changes that take place during the binding process of drug to HSA [6]. Up to now, the regions of binding best defined in HSA molecule are the sites warfarin-azapropazone (site I) and indol-benzodiazepine (site II) [7-11].

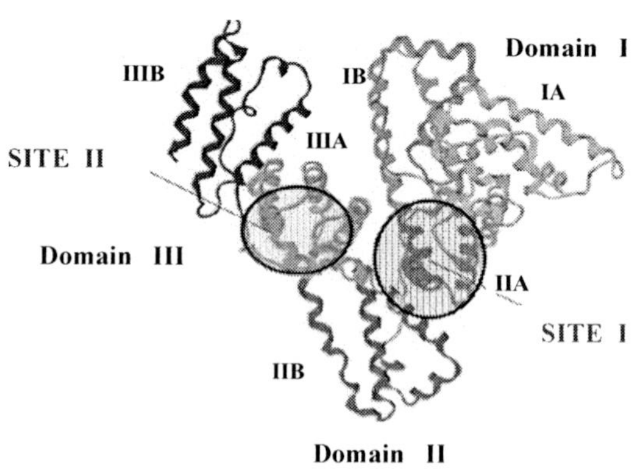

Figure 1. Structure of HSA. Domain I (red); domain II (green); domain III (blue).

The site I is in the sub-domain IIA and is formed by six helices that make up the sub-domain and a helix bound to the sub-domain IA [12] (Figure 1); the interior of the cavity is predominantly apolar though it contains two groups of polar residues, one at the end and the other one in the entry of the cavity. The site I seems to be specific for anionic or electronegative drugs such as warfarin, fenilbutazone and valproic acid. The interactions are hydrophobic and hydrogen bonds with the hydroxyl group of the polar group Y150 [12].

The site II is in the sub-domain IIIA; it is apolar and presents major selectivity than the site I for benzodiazepines and their derivatives, as well as for different carboxylic acids such as NSAIDs.

In spite of sites I and II being responsible for a great number of interactions between drugs and HSA, some crystalographic [13,14] and chromatographic studies [14,15] confirm the existence of other regions of binding in the HSA molecule of minor importance such as the binding site of digitoxine (site III) and the binding sites of billirubin and of tamoxifen, where red phenol and cis/trans clomiphen bind, respectively. Nowadays, the location of the binding sites of tamoxifen and billirubin and site III are not exactly known; but there are evidences that confirm that the site III is far from sites I and II, that there are allosteric effects between the site I and the binding site of tamoxifen, and that the site of billirubin and tamoxifen are overlapped.

AGP is a α_1-globulin protein that is characterized by its high content in carbohydrates. It is the plasmatic protein of minor molecular mass (41000 g/mol) and its content in sialic acid awards an isoelectric point of 2.7 [16]. Its plasmatic levels are between 0.4–1g/L, with a plasmatic concentration of 20 μM.

The polypeptidic chain of AGP is formed by 183 amino acids and has a content in carbohydrates of 41–45 %. AGP has three polymorphic variants (F1, S and A) with different primary structure. Its three-dimensional structure and physiological function are still not solved [17]. AGP presents a selective binding site and other sites with low affinity [4], up to the moment 7 sites have been found [18]. AGP has affinity for basic drugs such as tricyclic antidepressants, local anesthesics, phenothiazines and β-blockers and in minor extent, for some neutral and acid drugs. In general, the drug-AGP interactions are of hydrophobic nature.

Lipoproteins are spherical pseudo-micellar particles soluble in water with a hydrophobic nucleus composed principally of triglycerides, esters of cholesterol, sphingolipids and A, D, E and K vitamins surrounded with a hydrophilic layer of phospholipids, non esterified cholesterol and apoproteins. Lipoproteins are divided into four groups: kilomicrons, very low density lipoproteins (VLDL), low density lipoproteins (LDL) and high density lipoproteins proteins (HDL). These proteins principally bind to liposoluble drugs, with a high volume of distribution and generally of basic nature such as imipramine and cyclosporine [19].

The term globulin refers to all plasmatic proteins except HSA and prealbumin. There are principally five types of globulins: α_1-, α_2-, β_1-, β_2- and γ- globulin. α- and β- globulins have great affinity for different endogenous and exogenous substances of similar structure such as steroids (prednisone and transcortin) whereas γ- globulins specifically interact with antigens but their interaction with the majority of drugs is inappreciable [19].

2. Binding of Drugs to Plasmatic Proteins

The study of the interaction of drugs with plasmatic proteins is of utmost importance for the understanding of the pharmacokinetic and pharmacodynamic properties of drugs. Pharmacokinetics studies the absorption, distribution, metabolism and excretion (ADME) of a drug by the organism. On the other hand, pharmacodynamics is devoted to the study of the effects of a drug in the organism, such as the release of certain endogenous substances, the binding to a receptor in order to reduce or increase a certain stimulus or response, or provoking a toxic reaction [20].

From the pharmacokinetic point of view, the binding of a drug to plasmatic proteins can determine its distribution and elimination. Regarding distribution, when a drug is administered, it is absorbed and reaches the systemic circulation where it is distributed into the blood cells, plasmatic proteins and plasmatic water. Only the free fraction of drug is capable to reach the different organs and tissues. The bound fraction of drug acts as a reservoir since it can not be excreted, metabolized or distributed to the tissues. Concerning elimination, only the free fraction of drug can suffer renal or liver clearance [19]. Regarding pharmacodynamics, only the free drug has therapeutic or toxic action since it is the only one capable of crossing barriers and distribute into the tissues.

The interactions of drugs with plasmatic proteins are selective and of diverse nature. Principally, interactions are reversible physical bindings such as hydrogen bonding or Van der Waals forces, which can have a high degree of selectivity even can be enantioselective for some compounds [21]. Hydrophobic interactions are less selective and reversible. Irreversible covalent bonding of drugs to plasmatic proteins rarely occurs [2]. This irreversible bonding is responsible for certain types of toxicity such as the carcinogenic effect of chemical substances.

3. Methods to Evaluate Drug-Plasmatic Proteins Interactions

Nowadays, there is a great variety of biochemical techniques to evaluate the affinity of drugs to plasmatic proteins. The most used techniques are based on the separation of the free fraction of drug once the equilibrium has been reached, by means of equilibrium dialysis [22], ultrafiltration [23], ultracentrifugation [22] and gel filtration [24], followed by the determination of the free drug, which can be a problem in case of drugs with a high affinity towards proteins. Other disadvantages such as lack of automatization, long analysis time, non-specific drug adsorption onto devices or membranes, Donnan effect, sieve effect and high sample volumes, have been reported [25].

Spectroscopic techniques such as nuclear magnetic resonance (NMR) [26], molecular fluorescence [27] and circular dicroism [24] provide useful information about the binding of drugs to the proteins but the analysis time is long. Recently, optical biosensors such as IAsys have shown to be useful in the monitoring of the drug-HSA binding, providing information about the kinetics of the interactions and the bound drug concentration [28]. In the IAsys technology, the protein is immobilized by covalent binding onto the surface of the chip of the biosensor and after the addition of a drug solution, the association process is monitorized; if a buffer is used instead of the drug solution the dissociation process can be monitorized. The binding of drug to the protein provokes a change in the refraction index of the sensor surface increasing the signal, which is directly related to the amount of drug bound to the protein. This methodology has shown to very useful in the screening of drugs to asses the affinity of drugs toward proteins. It presents some advantages over the conventional techniques such as: (i) Low consumption of drug and protein, (ii) short analysis time (4-5 min for the characterization of the interaction, 20-25 min for the immobilization of the protein); (iii) simultaneity in the analysis and results acquisition; (iv) reutilization of the sensor chip, since the immobilized protein is stable for 1 month and (v) low cost.

In the last years, chromatographic and electrophoretic techniques have been widely used. In these techniques proteins can be used in solution (in the mobile phase or in the electrophoretic buffer) or immobilized onto the capillary or the stationary phase. The use of immobilized proteins in chromatographic and electrophoretic techniques provides lower protein consumption compared with the use of protein solutions. However, it presents some disadvantages such as (i) changes in the spatial arrangement or conformation of the binding sites of the protein with respect to the ones in plasma, and consequently, variation of the binding properties of the protein [29], (ii) lack of reproducibility if the immobilization is home-made (iii) complex methodology and (iv) low stability of the commercial capillaries and columns. So, if possible (i.e. protein availability), methods with protein in solution are preferable.

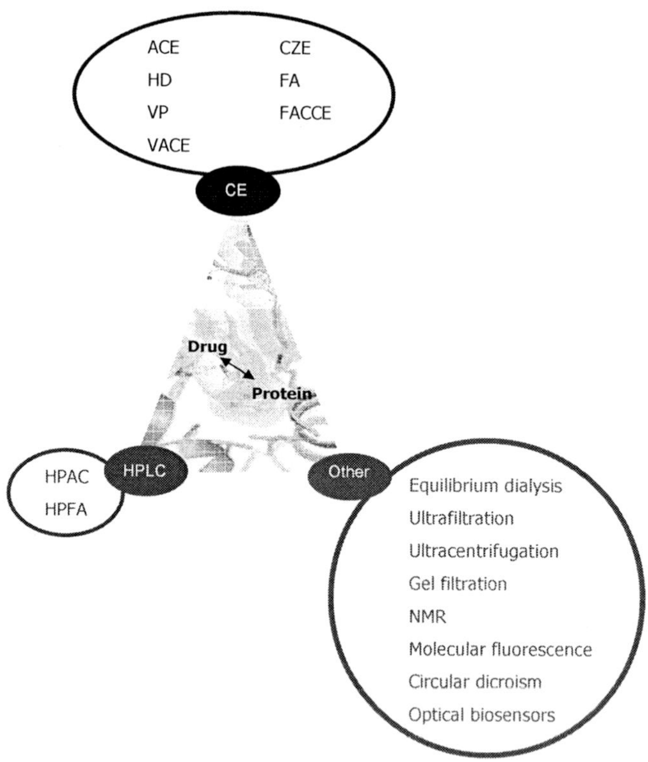

Figure 2. Analytical techniques for the evaluation of the drug-plasmatic proteins interactions.

Figure 2 shows a scheme with all the analytical technologies used for the evaluation of the binding of drugs to the plasmatic proteins.

3.1. Chromatographic Techniques

The chromatographic techniques used for the study of the affinity of drugs towards proteins are *high performance affinity chromatography (HPAC) and high performance frontal analysis (HPFA).* These techniques present several advantages over the classical biochemical techniques (ultrafiltration, ultracentrifugation, equilibrium dialysis, etc...). Besides the speed, precision and versatility inherent to the chromatographic techniques, the adsorption of drug into membranes or the pass of drug across them are avoided.

In *HPAC*, the protein under study, HSA or AGP, is immobilized onto the stationary phase [30]. Some advantages of this methodology for the evaluation of the drug-protein interactions are: (i) the possibility of re-using the column; i.e. HSA columns can be used for 500-1000 injections [31]; (ii) easy automation and (iii) short analysis time (5-15 min).

In HPAC there are two working methodologies: the zonal and frontal elution. The zonal elution consists in injecting a small quantity of sample that contains the drug under study into the column. The chromatogram shows a peak that corresponds to the drug, whose retention is directly related to the quantity of drug bound to the immobilized protein. By means of the injection into the chromatographic system of a series of drug samples of increasing concentration, it is possible to estimate the number of binding sites of the drug in the protein molecule and the corresponding affinity constants to these sites. By means of this methodology the affinity constants of benzodiazepines, coumarines, warfarin [32], antibacterial, anti-retroviral and anti-inflammatory drugs [33] to HSA have been estimated and studies of competitive binding for the determination of the location and structure of the binding sites in HSA molecule have been carried out [15].

In the frontal elution a sample that contains the drug under study is continuously injected into the column. First, the drug binds to the protein until the column is saturated and at this point drug is gradually eluted. The chromatogram shows a plateau and the time required to display this plateau depends on the total concentration of injected drug, the quantity of protein in the column and the drug-protein affinity constant. Frontal analysis needs larger amount of drug than zonal elution. This method also allows to obtain simultaneously the number of binding sites and the affinity constants. To determine both parameters, a series of samples containing drug at increasing concentrations are injected and the plateau times are measured. Using this method, the affinity constants and the number of binding sites of warfarin and tryptophan to HSA have been determined [34].

In *HPFA* a great volume of pre-equilibrated sample that contains drug and protein is injected into a chromatographic column with a stationary phase that hydrophobically retains small molecules of drug but not the proteins. During the chromatographic process, the dissociation of the drug-protein complex takes place. As result, proteins are first eluted and later the liberated drug as a plateau. The height of the plateau is related to the concentration of liberated drug and to the concentration of free drug in the solution of sample, so the concentration of free drug can be calculated from the height of the plateau. This methodology is of great interest for drugs showing high affinity for the protein due to the liberation of the bound drug into the column [35]. In addition, HPFA has been coupled to a preconcentration column and to a chiral column on line for the enantioselective racemic drug-protein binding studies.

3.2. Electrophoretic Techniques

In the last years, capillary electrophoresis (CE) has shown to be an attractive analytical tool for the study of drug- plasmatic proteins interactions due to its speed, high efficiency, great resolution power, rapid optimization and low consumption of reagents and samples.

The most widely used electrophoretic methodologies are: affinity capillary electrophoresis (ACE) [36, 37], Hummel-Dreyer's method (HD), vacancy peak method (VP), vacancy affinity capillary electrophoresis (VACE) [38], capillary zone electrophoresis (CZE) [38], frontal analysis (FA) [4, 40-42] and frontal analysis-continuous capillary electrophoresis (FACCE) [37,42]. These methodologies are used for the individual study of the interaction of drugs with HSA, AGP, lipoproteins and globulins. However, it is not possible to use directly plasma samples without elimination of particles in suspension that can provoke problems in the separation system. With the exception of VACE, these methodologies were previously developed for HPLC and then adapted to CE [37]. Table 1 summarizes the sample and electrophoretic solution composition as well as the variable used for the determination of the affinity constant of the drug-protein affinity constants in each electrophoretic method.

Table 1. Electrophoretic methods for the estimation of the drug-plasmatic protein affinity parameters.

Method	Sample	Electrophoretic solution	Analytical signal
ACE	D	B+P	Change in the mobility of D
	P	B+ D	Change in the mobility of P
HD	B+D+P	B+P (or D)	Vacancy peak area related to the DP concentration
VP	B	B+D+P	Vacancy peak area related to the free D concentration
VACE	B	B+D+P	Change in the mobility of D or P
ZCE	B+D+P	B	Peak area related to the free D concentration
FA	B+D+P	B	Plateau height related to the free D concentration
FACCE	B+D+P	B	Plateau height related to the free D concentration

B: electrophoretic buffer
D: drug
P: protein
DP: drug-protein complex

Affinity capillary electrophoresis (ACE) consists in filling the capillary with the electrophoretic buffer that contains protein and injecting the drug studied. To estimate the affinity constant, diverse experiences are carried out at constant drug concentration and increasing the protein concentration in the electrophoretic buffer, so the change in the electrophoretic mobility of the drug is related to the protein concentration [37]. Similarly, experiences injecting a constant protein concentration and filling the capillary with electrophoretic buffer containing increasing drug concentrations can be carried out. This second modality is not widely used due to the changes in the electrophoretic mobility of the drug-protein complexes are small since its molecular mass is high and similar to that of the protein. On the contrary, the increase of the protein concentration in the electrophoretic buffer provokes important changes in the mobility of the drug and the estimated affinity constants present a major precision. For the estimation of the affinity constants in ACE some considerations should be taken into account:

- The electroosmotic flow marker does not have to interact neither with the capillary walls nor with proteins.
- The interaction of the drug and the protein with the capillary wall must be null.
- The electrophoretic mobility of the drug-protein complex is considered to be approximately equal to the mobility of the drug at the major protein concentration studied.
- Kinetics of the drug-protein interactions must be quick in comparison with the migration time in order to reach the equilibrium.
- Due to sensitivity problems, it is impossible to work at therapeutic drug concentrations lower than μM.
- The drug must bind to a unique binding site type in the protein molecule.
- The electrical field must not alter the affinity of the drug towards the protein.

Using this methodology, the affinity constants of bilirrubine [43] and porphirine [44] to HSA and disopyramide to AGP [45] have been determined. ACE has also been used for the analysis of enantiomers and the estimation of their affinity constants using ciclodextrins as chiral selectors [46]. Whereas that in ACE only a binding site in the protein molecule is considered, in the rest of electrophoretic methodologies, HD, VP, VACE, FA and FACCE, the number of independent binding sites and the corresponding affinity constants can be estimated.

The *Hummel-Dreyer's method (HD)* consists in carrying out a series of experiences by filling the capillary with the electrophoretic buffer that contains a fixed drug concentration and injecting samples constituted by a fixed protein concentration and increasing drug concentrations. When the drug and protein concentrations in the sample are null, the electropherogram shows a negative peak. By injecting samples containing protein and increasing drug concentrations a positive peak appears in the electropherogram related to the drug-protein complex; the area and height of the negative peak decrease being positive at high drug concentrations. By using internal or external calibration the concentration of bound drug is determined when the negative peak disappears [5, 38]. Furthermore, some experiences in which the electrophoretic buffer contains a fixed protein concentration and injecting samples containing a fixed drug concentration and increasing protein concentrations can be carried out. In this case the area and height of the negative peak are related to the bound drug-protein concentration.

In the *vacancy peak method (VP),* the capillary is filled with the electrophoretic buffer that contains drug and protein and a small quantity of buffer (without drug and protein) is injected. Series of experiences are performed keeping constant the drug or protein concentrations and varying the concentration of the other compound. The electropherogram shows two negative peaks, one of them related to the drug-protein complex and the free protein concentration and the other peak is related to the free drug. The concentration of free drug in equilibrium with the protein is measured at each assayed concentration.

In *vacancy affinity capillary electrophoresis (VACE)* methodology, the experimental design is similar to that in VP but to estimate the affinity of the drug for the protein the change in the electrophoretic mobility of the drug or the protein is used.

In *capillary zone electrophoresis (CZE)*, pre-equilibrated samples containing drug and protein mixtures at different concentrations are injected and the separation of the free drug from the bound drug fractions is carried out. This methodology is only valid for those systems with high affinity constants and slow dissociation kinetics [39].

In *frontal analysis (FA)* method, the capillary is filled with the electrophoretic buffer and a considerable volume of pre-equilibrated sample (100-200 nL) that contains free drug, free protein and drug-protein complex is injected. To apply this methodology it is considered that the complex and the free protein have the same electrophoretic mobility and that the mobility of the free drug sufficiently differs from that of the complex. On the other hand, with the injection of large sample volumes the zone of free drug separates only partially of the zone of protein and complex, assuring the equilibrium state during all the electrophoretic development.

A series of experiences by injecting samples that contain buffer, a fixed concentration of drug and increasing concentration of protein (up to physiological concentration) are carried out. The electropherogram shows two plateaus, one of them related to the free protein and the complex fractions and the other plateau is related to the free drug fraction. To estimate the affinity parameters, the free drug concentration at each total protein concentration is calculated by using the free drug plateau height. It is also possible to maintain the concentration of protein constant and to increase the concentration of drug, but this second alternative implies a higher consumption of protein.

In the *frontal analysis-continuous capillary electrophoresis (FACCE)* methodology, the capillary is filled with the electrophoretic buffer that contains a fixed protein concentration and the sample that contains a pre-equilibrated mixture of drug and protein is hydrodinamically injected. When a voltage is applied, the inlet vial contains the sample and the outlet vial, the electrophoretic buffer so, the sample is continuously and electrokinetically injected into the capillary at the same time that the separation process is carried out. In this methodology it is necessary a close migration between the drug-protein complex and the free drug to avoid the equilibrium dissociation [42].

To estimate the drug-protein interaction parameters, a series of samples that contain a constant protein concentration and increasing drug concentrations are prepared. If the mobility of the drug is higher than that of the protein, the electropherogram shows a first plateau corresponding to the elution of the free drug and a second plateau related to the free protein and the drug-protein complex. The parameter to be measured is the height of the first plateau corresponding to free drug.

Among the seven electrophoretic technologies used for the estimation of the protein binding of drugs to the plasmatic proteins, FA seems to be the most preferable methodology for diverse reasons. Firstly, in ACE and VACE a systematic desviation in the estimated value of the association constants can exist when the free concentration in the electrophoretic buffer is used instead of the real free drug concentration in the migration zones. Secondly, in HD and VP, the calibration process is very laborious and in FACCE, the consumption of sample is high. On the contrary, FA is simple, robust, it can be used in situations of multiple equilibrium and the sample consumption is lower than in the other methodologies. Nevertheless, compared with FA, FACCE offers lower detection limits and there are not problems related to slow binding kinetics.

As has been previously commented, to evaluate the binding of drugs to human plasma a previous separation technique is required before the injection of the sample in the electrophoretic system in order to avoid problems related to the presence of particles in the electrophoretic separation system. A solution to this problem is the ultrafiltration of the pre-equilibrated samples of drug and human plasma. In this technique only the free fraction of drug crosses the membrane of the separation system whereas plasmatic proteins and the drug-protein complex remain retained in the membrane due to their high molecular weight. So, the ultrafiltrate can be directly injected into the electrophoretic system in the modality of CZE.

4. Improvements in Frontal Analysis-CE

In the last years, a high number of articles about the application of the frontal analysis in capillary electrophoresis for the study of the drug-protein interactions have appeared in the bibliography [4, 41, 47, 48] since this methodology presents several advantages over the other electrophoretic methodologies such as experimental simplicity, robustness, speed, low consumption of reagents and it allows to obtain great information about the drug-proteins affinity. Nevertheless, FA presents some problems that restrict in certain way its employment as a screening method of this type of interactions. Firstly, in spite of being the most rapid method, the analysis time ranges between 10-15 minutes, time that can be excessive in high-throughput purposes. In addition, the conventional treatment of the experimental data to obtain the affinity parameters uses experimental variables (concentration of free and bound drug) in both coordinate axes, so a small experimental error can lead to erroneous results. To overcome these problems, alternative methodological and mathematical data treatments have been proposed.

As regards to the analysis time in frontal analysis, the short-end injection methodology has been proposed [40]. In this methodology vacuum hydrodynamic sample injection into the short-end of the capillary together with the application of a negative running voltage (inverse polarity) is used. The affinity parameters of the systems warfarin-HSA, alprenolol-HSA and alprenolol-AGP obtained using either long- or short-end injection into the capillary were similar but short-end injection methodology reduced the analysis time from 10-20 min to 1-4 min [40].

On the other hand, an equation for the estimation of the affinity parameters of drugs to plasmatic proteins that isolates experimental errors in one axis has been proposed [40]. The mathematical treatment of FA-CE data is based on the following assumptions.

When a certain drug solution is equilibrated with protein, the bound drug concentration ($[D]_b$) can be expressed as the difference between the total drug concentration (C_D) and the free drug concentration ($[D]_f$):

$$[D]_b = C_D - [D]_f \qquad \text{Eq. 1}$$

Considering that m types of independent binding sites per protein molecule exist and that drug binds to any of these sites with 1:1 stochoimetry, the total bound drug concentration can be expressed as:

$$[D]_b = \sum_{i=1}^{m} [D]_{b,i} = C_P \sum_{i=1}^{m} \frac{n_i K_i [D]_f}{1 + K_i [D]_f} \quad \text{Eq. 2}$$

being $[D]_{b,i}$, the concentration of bound drug at the site i, C_P, the total protein concentration, n_i and K_i are the number of binding sites and the affinity constant for each site, respectively. The product n_iC_P represents the total concentration of i-class binding sites. Equation 2 is the basis of the traditional data treatment for determining the binding parameters of drugs to proteins from the non-linear fitting of Equation 3:

$$r = \frac{[D]_b}{C_P} = \sum_{i=1}^{m} \frac{n_i K_i [D]_f}{1 + K_i [D]_f} \quad \text{Eq. 3}$$

where r is the concentration of bound drug molecules per protein molecule. Substituting Equation 1 into 3 and rearranging the terms it is possible to obtain the following expression that allows the estimation of the binding parameters by non-linear regression assuming that one or two types of independent binding sites (m=1 or m= 2) in the protein molecule are the main responsible of drug binding [41]. Therefore Equation 3 can be expressed as follows:

$$r = \frac{C_D - [D]_f}{C_P} = \left(\frac{n_1 K_1 [D]_f}{1 + K_1 [D]_f} + \frac{n_2 K_2 [D]_f}{1 + K_2 [D]_f} \right) \quad \text{Eq. 4}$$

In the most general case of one type of binding site (m=1), different linearized forms of equation 4, such as the Scatchard and the Klotz plots have been developed and reviewed recently [39]. However, one disadvantage of Equation 4 and its linearized forms is that experimentally measured concentration of free drug and its derived corresponding r-values are present in both dependent and independent variable. Therefore, small experimental errors are included in both axes leading sometimes to difficulties in the interpretation of results and to wrong estimations of the binding parameters.

One possible alternative to overcome this problem was proposed by McDonnell et al. [4, 49]. The authors proposed an equation relating the experimentally measured percentage of drug bound to the protein with the total drug concentration. However, the application of this equation is limited to a single class of interaction site (m= 1) assuming one binding site (n=1).

A more general treatment can be obtained from Equation 4 (m=1, n≠1). This assumption is based on the fact that the usual K_1 and K_2 values for proteins such as HSA are in the range between 10^5-10^4 and 10^3-10^2 M^{-1}, respectively. Therefore the bound drug concentration at the secondary binding site (m=2) under physiological conditions would only be appreciable at free drug concentrations over 10^{-3} M, while the usual therapeutic levels are very much lower (10^{-7}-10^{-4} M) [50]. In the case of AGP, drugs only have affinity for one type of binding site in this molecule so it could also be considered m=1. Therefore the second term between brackets in Equation 4 can be neglected in most cases. This leads to a simplification of Equation 4 giving the following expression:

$$[D]_f = C_D - C_P\left(\frac{n_1 K_1 [D]_f}{1 + K_1 [D]_f}\right) \qquad \text{Eq. 5}$$

Rearranging the terms of this last equation a second degree polynomial equation whose resolution provides Equation 6 that relates the free drug concentration $[D]_f$ with the total protein (C_P) and drug (C_D) concentrations [40].

$$[D]_f = \frac{-(1 - K_1 C_D + n_1 K_1 C_P) \pm \sqrt{(1 - K_1 C_D + n_1 K_1 C_P)^2 + 4 K_1 C_D}}{2K_1} \qquad \text{Eq. 6}$$

A non-linear plot of the experimentally measured $[D]_f$ vs. C_P or C_D can be applied to the data analysis being the number of primary binding sites (n_1) and its corresponding constant (K_1) the fitting parameters.

In order to estimate drug-protein interactions at near-physiological conditions, it is possible to work with two experimental procedures: (A) series with increasing total concentration of drug (C_D,) and a fixed total concentration of protein (C_P, approximately at the physiological protein concentration); or (B) series keeping constant C_D (close to the therapeutic level) and increasing C_P (from 0 up to physiological concentration). The (A) experimental approach would result in a great consumption of protein that turns into a higher cost per analysis. However the experimental procedure (B) allows reaching the physiological protein concentration with a much lower cost.

It is worth mentioning that reaching physiological protein concentration is a requirement to obtain realistic estimations of drug-protein interactions since it has been reported that displacement of bound drug may happens due to albumin aggregation at high protein concentration [38].

On the other hand, due to the lack of sensitivity inherent to the CE, total drug concentration below 10^{-4} M, leading to free drug concentrations between 10^{-4} and 10^{-6} M, are difficult to use in this kind of studies.

The protein binding percentages of drugs (PB) to proteins can be defined as the ratio between the bound drug concentration and the total drug concentration:

$$PB(\%) = 100\frac{C_D - [D]_f}{C_D} \qquad \text{Eq. 7}$$

4. Applications of Frontal Analysis-Capillary Electrophoresis to the Evaluation of Drugs-Plasmatic Proteins Interactions

Our research group has determined the affinity parameters of 50 compounds: 13 β-blockers, 4 phenothiazines, 17 antihistamines, 3 barbiturates, 3 benzodiazepines, and 10 polyphenolic compounds to HSA and AGP at near physiological conditions (T = 36.5°C, phosphate buffer at pH 7.4) by means of FA–CE [51-53]. For this purpose, mixtures containing a fixed drug concentration (C_D= 100-200 μM) and increasing concentration of HSA (C_P from 0 to 475 μM) or AGP (C_P from 0 to 20 μM) were prepared in duplicate. The free drug concentration at each total protein concentration was determined by the plateau height ratio between each sample and a sample containing only drug measured in the electropherograms.

To estimate the drug-proteins affinity constants, K_1, and the number of primary binding sites, n_1, the free drug concentrations, $[D]_f$, obtained for each protein concentration, C_P, and drug concentration, C_D, were adjusted to Eq. 6. Figure 3 shows as an example the experimental data and the binding curves of some compounds to HSA and AGP. Table 2 shows the estimated n_1 and K_1 values with the corresponding standard error for the drugs studied. For the compounds that did not interact with proteins a value of n_1 and K_1 of zero was assigned. For some compounds, although interactions with proteins exist, the experimental data could not be adjusted to equation 6 due to mathematical algorithm performance problems.

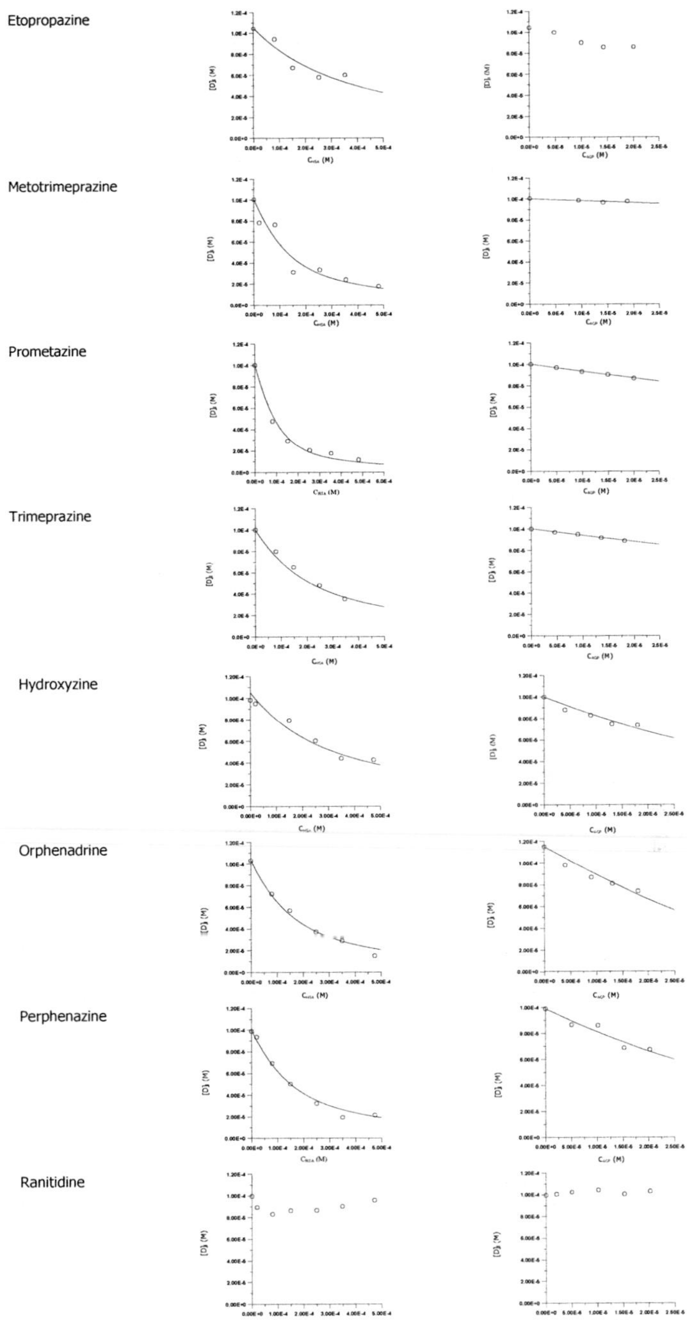

Figure 3. Binding curves to HSA (left side) and AGP (right side).

Table 2. Affinity constants, K_1, and number of primary binding sites, n_1, experimentally obtained [51-53]. The n_1 and values are expressed as the mean ± standard error

Drug	n_{1HSA}	K_{1HSA} (M^{-1})	n_{1AGP}	K_{1AGP}(M^{-1})
β-blockers				
Acebutolol	-	-	0	0
Alprenolol	1.00±0.09	$(3.7±0.5)·10^2$	1.00±0.07	$(4.3±1.3)·10^4$
Atenolol	0	0	0	0
Carteolol	0	0	1.00±0.12	$(2.4±0.9)·10^4$
Celiprolol	-	-	0	0
Labetalol	1.00±0.10	$(5.2±0.3)·10^2$	1.00±0.06	$(2.2±0.4)·10^4$
Metoprolol	0	0	0	0
Nadolol	0	0	-	-
Oxprenolol	0.9±0.3	$(4.6±0.2)·10^2$	1.00±0.13	$(3.5±1.1)·10^4$
Pindolol	1.03±0.3	$(1.3±0.4)·10^2$	1.2±0.11	$(3.0±0.2)·10^4$
Propranolol	0.82±0.2	$(1.09±0.08)·10^3$	1.00±0.15	$(4.6±1.0)·10^4$
Sotalol	1.1±0.3	$(1.5±0.3)·10^2$	0	0
Timolol	-	-	-	-
Phenothiazines				
Etopropazine	0.9±0.2	$(2.3±0.5)·10^3$	-	-
Metotrimeprazine	0.96±0.12	$(2.6±1.1)·10^4$	1.01±0.16	$(2.4±0.5)·10^3$
Prometazine	1.01±0.13	$(2.4±0.8)·10^4$	1.00±0.03	$(20.8±0.8)·10^3$
Trimeprazine	0.99±0.11	$(6.1±0.6)·10^3$	1.001±0.022	$(16.8±0.9)·10^3$
Antihistamines				
Brompheniramine	1	$(1.8±0.3)·10^3$	3.0±0.7	$(4.1±1.3)·10^3$
Carbinoxamine	-	-	3.0±0.4	$(25.0±0.6)·10^3$
Chlorcyclizine	1	$(5.2±1.1)·10^3$	-	-
Chlorpheniramine	1	$(3.7±0.7)·10^3$	2.0±0.3	$(16.1±0.7)·10^3$
Cinarizine	1	$(8.4±1.2)·10^3$	3.0±0.8	$(4.8±1.7)·10^3$
Cyclizine	1	$(1.29±0.10)·10^3$	-	-
Dimetindene	-	-	2.0±0.2	$(14.5±0.6)·10^3$
Doxylamine	1	$(7±1)·10^2$	2.0±0.2	$(1.73±0.12)·10^4$
Etintidine	0	0	0	0
Hydroxyzine	1	$(4.1±0.8)·10^4$	2.0±0.2	$(4.0±0.7)·10^4$
Orphenadrine	1	$(4.1±0.5)·10^3$	3.0±0.3	$(1.7±0.5)·10^4$
Perphenazine	1	$(9.5±0.8)·10^3$	3.02±0.12	$(6.4 ± 1.5)·10^4$
Phenindamine	1	$(10.2±0.7)·10^4$	3.0±0.3	$(1.8 ± 0.4)·10^4$
Ranitidine	0	0	0	0
Terfenadine	1	$(4.2±0.8)·10^3$	3.0±0.3	$(2.1 ± 0.6)·10^4$
Tripelenamine	1	$(17±3)·10^3$	1.0±0.3	$(4.3 ± 0.8)·10^4$
Tripolidine	1	$(7.5±1.4)·10^3$	2.0±0.3	$(1.0 ± 0.3)·10^4$

Table 2. (Continued)

Drug	n_{1HSA}	K_{1HSA} (M^{-1})	n_{1AGP}	K_{1AGP}(M^{-1})
Barbiturates				
Hexobarbital	1	$1.1\ 10^3$	1	$2.1\ 10^4$
Butabarbital	1	$1.2\ 10^3$	1	$5.3\ 10^3$
Mephobarbital	1	$1.1\ 10^4$	1	$3.7\ 10^4$
Benzodiazepines				
Oxazepam	1	$2.3\ 10^4$	1	$7.0\ 10^3$
Chlordiazepoxide	1	$6.3\ 10^4$	1	$7.0\ 10^4$
Diazepam	1	$7.1\ 10^4$	2	$1.4\ 10^4$
Polyphenolic compounds				
Chlorogenic acid	0	0	0	0
Apigenin	-	$>10^6$	0	0
Catechin	1	$(2.2\pm0.8)\cdot10^3$	-	-
Epicatechin	0	0	-	-
Flavanone	1	$(7.0\pm0.5)\cdot10^4$	0	0
Flavone	2	$(3.3\pm0.5)\cdot10^4$	0	0
Quercetin	1	$(5\pm3)\cdot10^5$	0	0
Rutin	1	$(6.9\pm1.5)\cdot10^3$	-	-
Vicenin-2	-	-	-	-
Vitexin	1	$(1.6\pm0.6)\cdot10^5$	0	0

(-) It could not be estimated because the experimental data did not fit toEq.6.

As can be observed in Table 2, the affinity constants of these compounds varied from 10^2 to 10^6 for HSA and from 10^3 to 10^4 for AGP; the number of primary binding sites in HSA and AGP molecules was 1 for most cases, except for antihistamines in the binding to AGP, being the number of sites 2 or 3.

To characterize the interactions of drugs with plasmatic proteins, the drug protein binding percentage values to HSA (PB_{HSA}), to AGP (PB_{AGP}) and to all plasmatic proteins ($PB_{whole\ plasma}$) in physiological conditions were determined using the following methodology. Four types of samples were prepared in duplicate containing: (i) 100-200 μM drug; (ii) 100-200 μM drug and 475 μM HSA, (iii) 100-200 μM drug and 20 μM AGP and (iv) 100-200 μM drug and 200 μL plasma. After that, samples were pre-equilibrated for 30 min at 36.5°C and injected into the electrophoretic system. The free drug concentration was determined by the plateau height ratio between sample containing HSA or AGP and the sample containing only drug. In the case of plasma solutions, samples were ultrafiltrated through regenerated cellulose membrane filter devices (with a MW cut-off of 10000 daltons) prior to the injection into the electrophoretic system. In this case, the free drug concentration was calculated by the peak area ratio between the ultrafiltrated and the solution containing only drug.

Table 3 shows the estimated protein binding percentages of the drugs studied to HSA, AGP and whole plasmatic proteins at a drug concentration of 100-200 μM. The bibliographic data available taken from different sources for the $PB_{whole\ plasma}$ have also been included, showing a good concordance with the experimental data [51-53]. Due to the sensitivity of the electrophoretic system, it was not possible to work at drug concentrations lower than 100-200 μM. Therefore; using the estimated values of K_1 and n_1 for each compound and the physiological concentrations of HSA and AGP, the protein binding percentage to both proteins at therapeutic drug concentrations were estimated by using equations 6 and 7 (C_{t1} and C_{t2} for HSA and AGP, respectively). As can be observed in Table 3, the estimated protein binding percentages to HSA and AGP were similar at 100-200 μM and the therapeutic drug concentrations for all the families studied except for β-blockers in their binding towards AGP. This effect can be explained by considering a saturation of the binding sites of AGP since β-blockers present a high affinity towards this protein and the concentration of drug is higher than the protein concentration (20 μM).

According to their affinity towards plasmatic proteins, the drugs studied were classified in 6 groups:

(1) Drugs that principally bind to HSA: acebutolol, sotalol, metotrimeprazine, prometazine, trimeprazine, brompheniramine, chlorcyclizine, cinarizine, cyclizine, orphenadrine, perphenazine, phenindamine, terfenadine, tripelenamine, tripolidine, hexobarbital, mephobarbital, butabarbital, oxazepam, diazepam, chlordiazepoxide, apigenin, quercetin, vitexin, flavanone and flavone.
(2) Drugs that principally bind to AGP: carbinoxamine, carteolol, alprenolol, pindolol and vicenin-2.
(3) Drugs that principally bind to HSA and AGP: labetalol, timolol, chlorpheniramine, doxylamine and hydroxyzine.
(4) Drugs that bind to lipoproteins and/or globulins: dimetindene, etintidine, metoprolol, ranitidine, epicatechin and chlorogenic acid.
(5) Drugs that bind to all plasmatic proteins: celiprolol, nadolol, oxprenolol, propranolol, etopropazine, catechin and rutin.
(6) Drugs that do not bind to proteins: atenolol.

Table 3. Protein binding percentages to plasmatic proteins

Drug	PB_{HSA} (%)		PB_{AGP} (%)		$PB_{whole\ plasma}$ (%)	
	Exp_1	C_{t1}	Exp_2	C_{t2}	Exp_3	Bib.
β-blockers						
Acebutolol	17±4	-	0	0	24±6	26±3
Alprenolol	20±4	20.2	17.85±0.07	49.4	66±2	76-85
Atenolol	3±1	0	0	0	0	<5
Carteolol	0	0	15±2	33.6	14.8±1.2	16
Celiprolol	10.1±0.9	-	3.4±0.2	0	29±4	26
Labetalol	23±2	29.4	14.7±1.3	28.9	59±3	50
Metoprolol	5±7	0	0	0	11.2±0.8	11±1
Nadolol	0	0	12.2±1.5	-	34±2	20-30
Oxprenolol	19.7±0.6	18.5	16±3	45.4	76±2	80-95
Pindolol	8±2	9.3	14.1±1.3	39.6	46±1	51±3
Propranolol	31±3	33.0	19.6±0.2	45.6	97±3	85-95
Sotalol	8±3	8.5	2.15±0.10	0	8.2±0.5	<5
Timolol	18±2	-	15.3±1.3	-	28±3	10,60±3
Phenothiazines						
Etopropazine	54±9	55.8	18±3	-	99±1	-
Metotrimeprazine	88±8	93.5	4±2	4.56	98.5±0.3	-
Prometazine	89±3	92.4	14.1±0.3	20.4	97±4	75-93
Trimeprazine	74±3	79.5	13.1±0.3	26.9	98±2	-
Antihistaminas						
Brompheniramine	44±5	49.7	17±4	21.3	90±10	-
Carbinoxamine	11±3	-	40.5±0.6	39.7	69±5	-
Chlorcyclizine	68±10	74.1	21±3	-	100±2	-
Chlorpheniramine	60±9	67.0	18±6	24.1	100±3	-
Cinarizine	77±7	82.2	19±6	19.0	100±2	-
Cyclizine	33±4	41.5	34±5	-	76±1	70±3
Dimetindene	13±2	-	22.9±0.5	23.2	83±2	-
Doxylamine	24±4	27.8	25±2	24.9	38±6	-
Etintidine	0	0	0	0	49±4	-
Hydroxyzine	96±11	95.7	27±2	32.2	100±2	-
Orphenadrine	63±7	69.3	35±4	34.7	81±10	93
Perphenazine	79±3	83.9	30±5	50.2	93±1	95
Phenindamine	80±3	84.6	36±4	35.5	100±2	-
Terfenadine	63±3	69.8	38±5	37.5	95±6	97
Tripelenamine	87±6	90.3	16.1±0.8	17.2	78±3	-
Tripolidine	75±8	80.5	20±3	19.6	88±1	-

Table 3. (Continued)

Drug	PB_{HSA} (%)		PB_{AGP} (%)		$PB_{whole\ plasma}$ (%)	
	Exp_1	C_{t1}	Exp_2	C_{t2}	Exp_3	Bib.
Barbiturates						
Hexobarbital	34.7	37.7	14.1	29.4	50.2±0.1	42-52
Butabarbital	38.6	39.7	6.6	9.6	24.5±0.5	<26
Mephobarbital	78.9	85.8	16.1	15.2	60.2±0.1	59-67
Benzodiazepines						
Diazepam	89.9	97.5	22.7	35.8	100	98.7±0.2
Chlordiazepoxide	99.8	97.2	20.5	17.1	100	94-97
Oxazepam	95.1	92.7	11.5	12.2	100	98.8±1.8
Polyphenolic compounds						
Chlorogenic acid	0±1	0	0	0	73±3	-
Apigenin	100±1	99.8	0	0	55±1	-
Catechin	45±10	54.7	21±7	-	90±3	-
Epicatechin	0±1	0	13±6	-	74±7	-
Flavanone	100±1	97.5	0	0	94±8	-
Flavone	98±1	97.3	0	0	100±1	-
Quercetin	93±1	99.6	0	0	92±11	-
Rutin	56±10	79.1	19±9	-	92±3	-
Vicenin-2	10±12	-	40±10	-	42±7	-
Vitexin	97±3	98.9	0	0	46±15	-

In order to determine the physicochemical and structural properties determining the interactions of drugs to plasmatic proteins, principal component analysis (PCA) were performed using as **X**-variables hydrophobic, steric and electronic parameters of these drugs together with their protein binding values to HSA, AGP and lipoproteins and/or globulins at therapeutic drug concentration. Since these variables were in different scales, data were autoscaled before performing the PCA. Table 4 shows the structure of the drugs and the structural descriptors used. Hydrophobic parameters includes the logarithm of octanol-water partition coefficients (log P) and retention data in biopartitioning micellar chromatography (log k_{BMC}); as electronic parameters the acidity constants (pKa) and the polar surface area (PSA) were used; finally molar volume (V_M), molar refractivity (R_M) and polarizability (Pol) were used as steric parameters. [51-53]

Figure 4 shows the loading plots corresponding to the first two principal components for the basic drugs β-blockers, phenothiazines and antihistamines

(Figure 4A) and for neutral and acidic drugs barbiturates, benzodiazepines and polyphenolic compounds (Figure 4B).

A

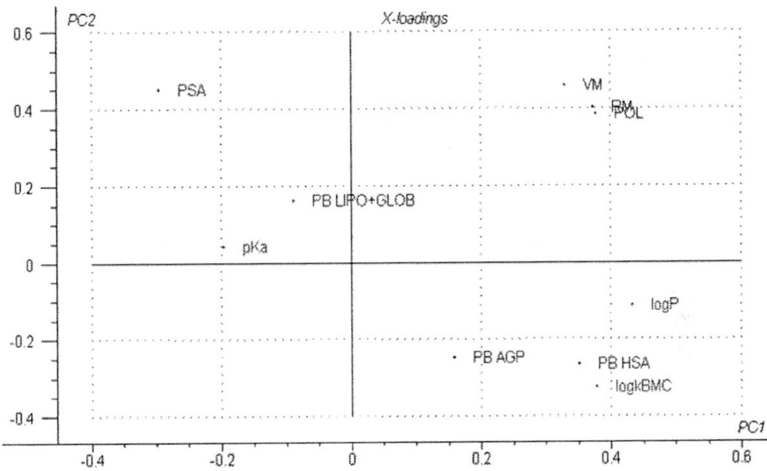

B

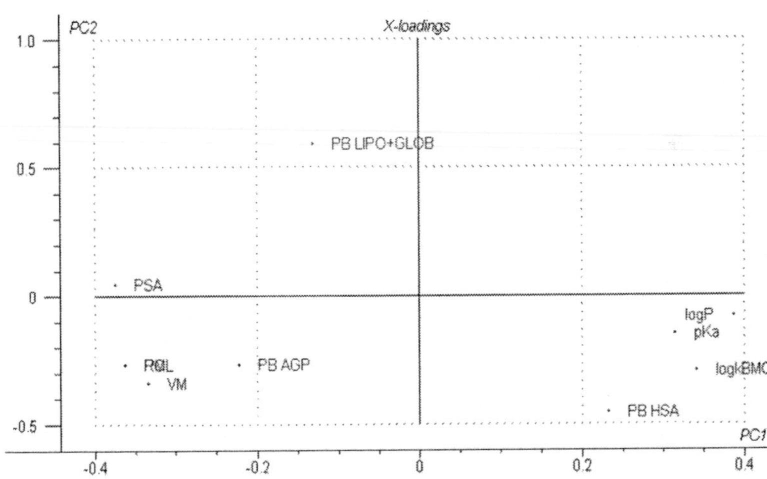

Figure 4. Loading plots for the first two principal components in the PCA analysis for: (A) β-blockers, phenothiazines and antihistamines; (B) barbiturates, benzodiazepines and polyphenolic compounds.

Table 4. Structure, physico-chemical and structural properties of β-blockers, phenotiazines, antihistamines, barbiturates, benzodiazepines and polyphenolic compounds.

Drug	Structure	pKa	LogP	$Logk_{BMC}$	V_M (cm^3)	R_M (cm^3)	Pol($\cdot10^{24}$) (cm^3)	PSA ($Å^2$)
β-blockers								
Acebutolol	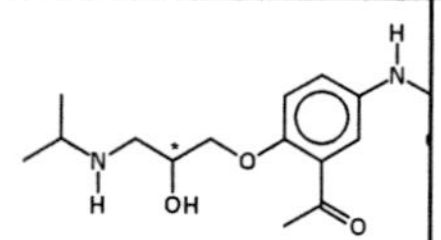	9.20	1.61	0.39	300.6	94.87	36.45	87.66
Alprenolol	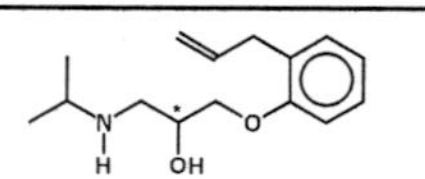	9.65	3.10	1.35	247.4	74.76	29.43	41.49
Atenolol	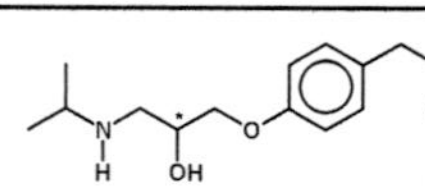	9.60	0.16	-0.22	236.6	73.51	29.03	89.16
Carteolol	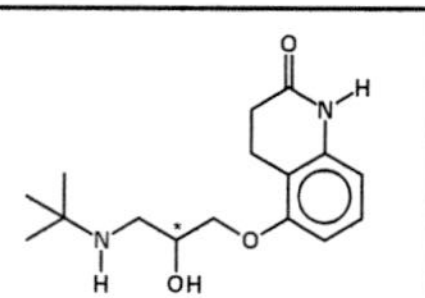	9.24	1.17	0.06	258.5	83.14	32.03	76.17
Celiprolol	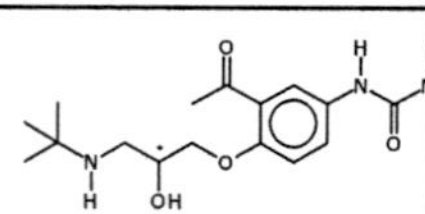	9.68	1.66	0.33	340.6	108.25	41.32	95.48

Table 4. (Continued)

Drug	Structure	pKa	LogP	$Logk_{BMC}$	V_M (cm^3)	R_M (cm^3)	Pol ($\cdot 10^{24}$) (cm^3)	PSA ($Å^2$)
Labetalol	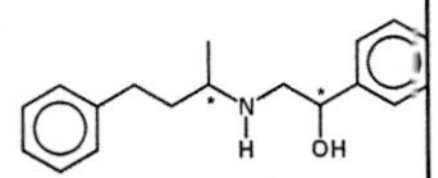	7.40 8.70	3.09	1.33	273.5	94.72	36.47	100.16
Metoprolol	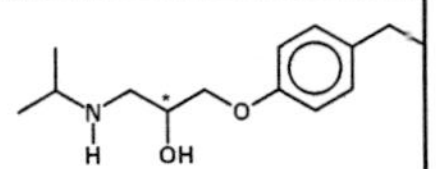	9.70	1.88	0.46	258.7	76.70	30.30	55.30
Nadolol	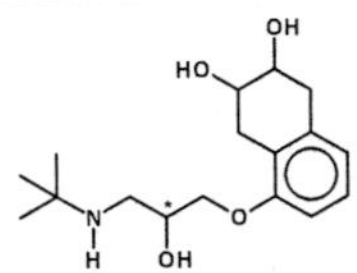	9.39	0.71	-0.08	260	85.53	33.76	86.53
Oxprenolol	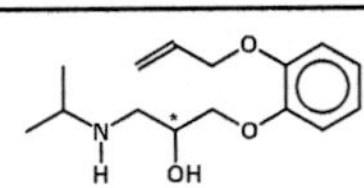	9.50	2.18	1.13	255.1	76.00	30.16	55.30
Pindolol	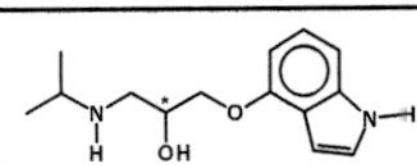	8.80 9.70	1.75	0.62	215.4	71.46	29.43	61.86
Propranolol	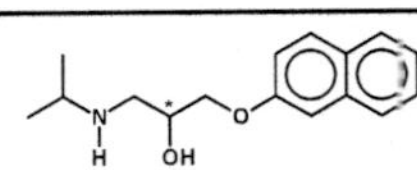	9.45	3.56	1.62	237.1	76.83	31.77	46.07
Sotalol	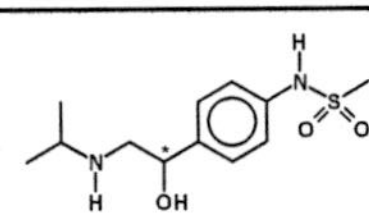	9.05	0.23	-0.1	219.7	71.12	28.83	91.39

Table 4. (Continued)

Drug	Structure	pKa	LogP	$Logk_{BMC}$	V_M (cm^3)	R_M (cm^3)	Pol ($\cdot 10^{24}$) (cm^3)	PSA ($Å^2$)
Timolol	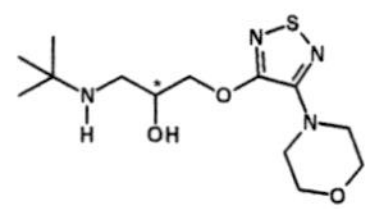	9.21	1.91	0.61	258.4	83.92	31.51	112.56
Phenotiazines								
Etopropazine	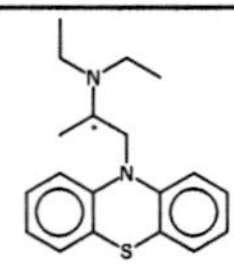	9.60	5.54	2.17	284.3	98.00	37.85	32.98
Metotrimeprazine	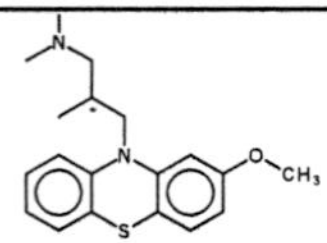	9.20	4.68	2.17	291.8	99.83	38.48	42.21
Prometazine	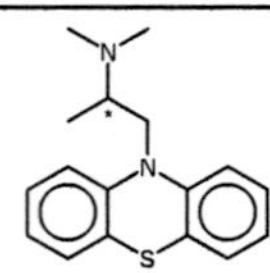	9.10	4.65	2.37	251.3	88.50	34.15	32.98
Trimeprazine	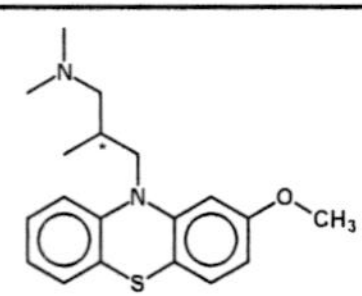	9.00	4.59	2.24	267.8	93.37	36.00	32.98

Table 4. (Continued)

Drug	Structure	pKa	LogP	$Logk_{BMC}$	V_M (cm^3)	R_M (cm^3)	Pol ($\cdot 10^{24}$) (cm^3)	PSA ($Å^2$)
Antihistamines								
Brompheniramine	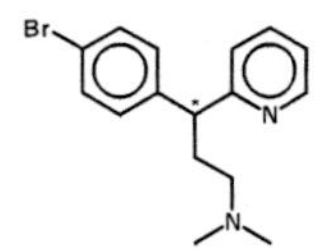	9.79	3.78	1.88	252.2	83.67	32.34	16.13
Carbinoxamine	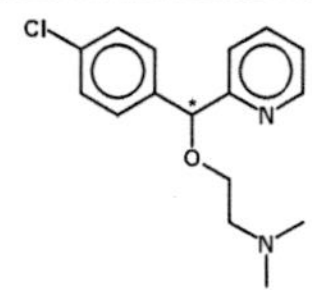	8.10	3.10	1.80	254.4	82.13	32.30	26.56
Chlorpheniramine	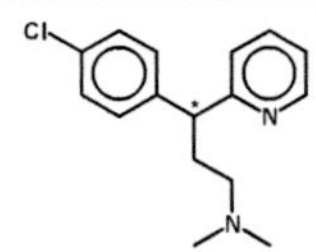	9.16	3.50	1.77	248.0	80.85	31.55	17.33
Cinarizine	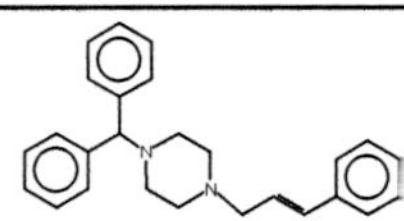	7.80	6.14	2.26	337.1	119.30	46.58	6.48
Cyclizine	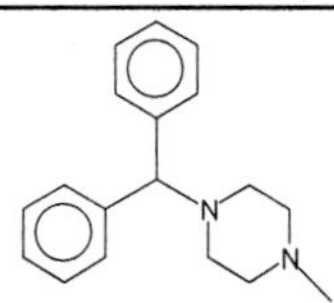	2.54 8.32	3.97	1.96	250.7	83.85	33.24	6.48

Table 4. (Continued)

Drug	Structure	pKa	LogP	$Logk_{BMC}$	V_M (cm^3)	R_M (cm^3)	Pol ($\cdot 10^{24}$) (cm^3)	PSA ($Å^2$)
Dimetindene	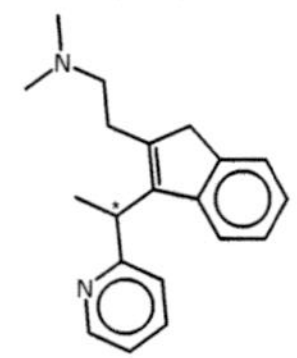	6.58	3.81	1.94	274.4	93.57	36.32	17.33
Doxylamine	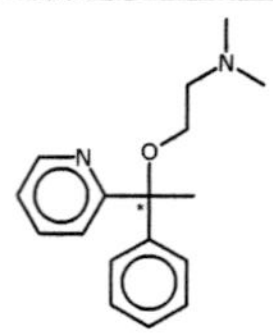	4.40 9.20	2.80	1.39	259.0	82.24	32.32	26.56
Etintidine	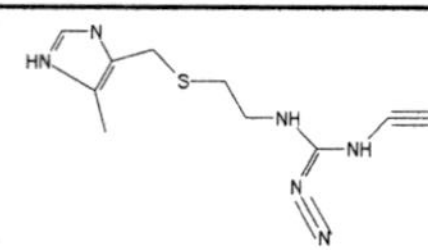	9.85	0.61	0.68	198.2	70.70	25.51	115.43
Hydroxyzine	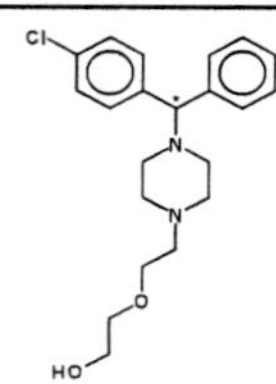	2.10 7.10	3.49	1.60	317.1	107.07	41.99	37.14

Table 4. (Continued)

Drug	Structure	pKa	LogP	$Logk_{BMC}$	V_M (cm^3)	R_M (cm^3)	Pol ($\cdot 10^{24}$) (cm^3)	PSA ($Å^2$)
Orphenadrine	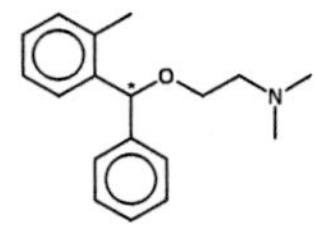	8.40	4.08	2.05	265.5	84.97	33.17	13.67
Perphenazine	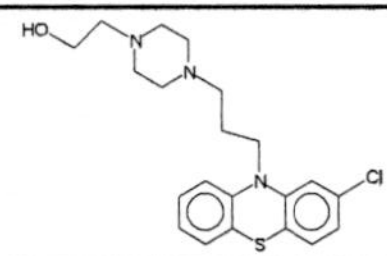	7.80	5.57	1.57	322.3	114.22	45.28	55.2
Phenindamine	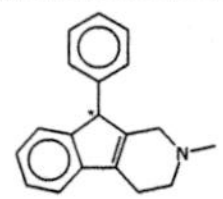	8.30	3.82	1.95	227.0	85.03	32.80	4.44
Ranitidine	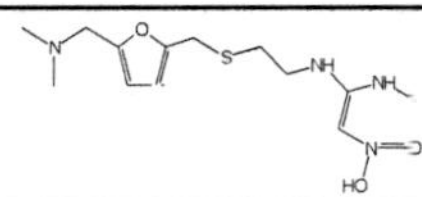	2.30 8.20	0.27	-0.15	265.4	85.64	33.95	106.05
Terfenadine	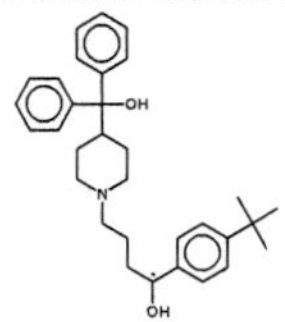	9.50	7.05	2.32	433.5	146.27	57.33	44.9
Tripelenamine	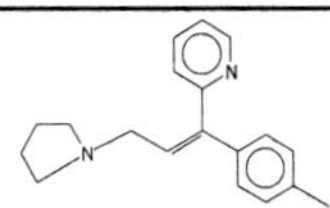	4.20 8.70	2.85	1.61	238.1	80.76	32.01	16.13

Table 4. (Continued)

Drug	Structure	pKa	LogP	$Logk_{BMC}$	V_M (cm^3)	R_M (cm^3)	Pol ($\cdot 10^{24}$) (cm^3)	PSA ($Å^2$)
Tripolidine	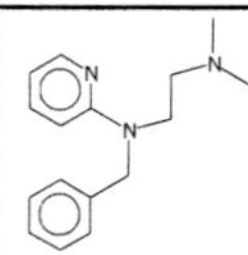	6.50 9.50	3.47	2.05	262.1	88.12	34.93	19.37
Barbiturates								
Butabarbital	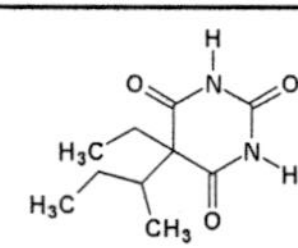	7.90	1.65	1.36	193.0	53.26	21.11	75.27
Hexobarbital	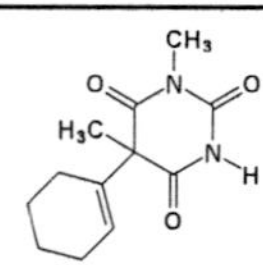	8.20	1.49	1.37	193.0	60.38	23.93	66.48
Mephobarbital	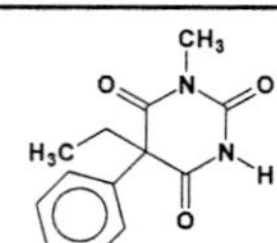	7.80	1.84	1.48	203.0	64.13	25.42	66.48
Benzodiazepines								
Chlordiazepoxide	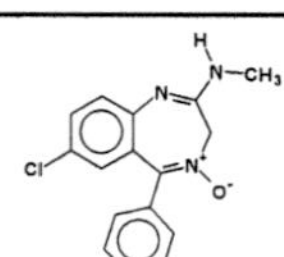	4.76	2.44	1.59	231.0	83.96	33.28	53.14

Table 4. (Continued)

Drug	Structure	pKa	LogP	$Logk_{BMC}$	V_M (cm^3)	R_M (cm^3)	Pol ($\cdot 10^{24}$) (cm^3)	PSA ($Å^2$)
Diazepam	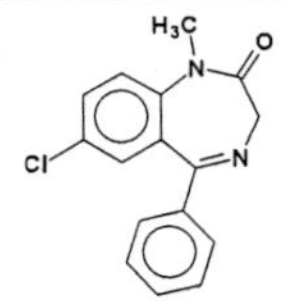	3.30	2.80	1.65	226.0	80.91	32.07	32.67
Oxazepam	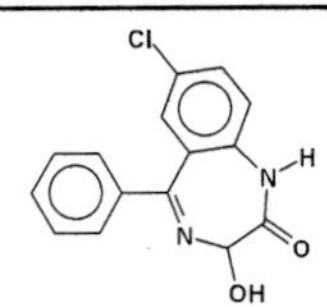	1.70 11.6	2.24	1.47	202.0	76.43	30.30	61.69
Polyphenolic compounds								
Chlorogenic acid	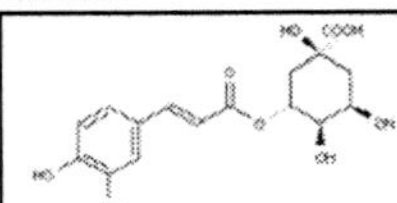	2.66	-1.01	-0.17	214.5	82.03	32.52	164.75
Apigenin	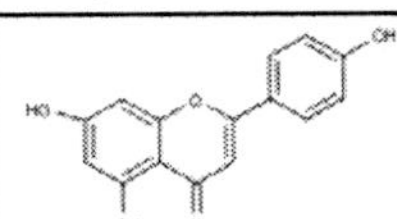	-	3.02	1.10	174.5	69.85	27.69	86.95
Catechin	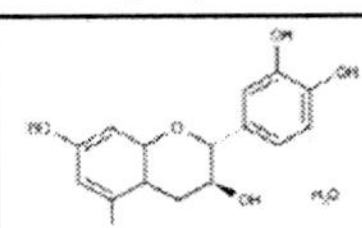	4.60	1.18	0.83	182.1	73.59	29.17	110.38

Table 4. (Continued)

Drug	Structure	pKa	LogP	$Logk_{BMC}$	V_M (cm^3)	R_M (cm^3)	Pol ($\cdot 10^{24}$) (cm^3)	PSA ($Å^2$)
Epicatechin	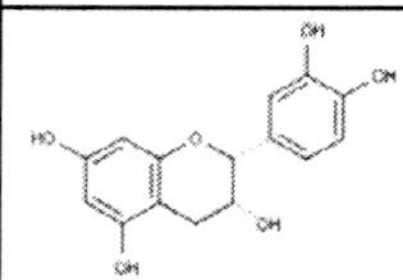	-	1.18	0.80	182.1	73.59	29.17	110.38
Flavanone	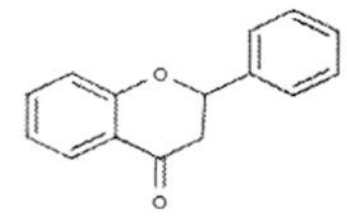	-	3.14	1.60	187.9	64.60	25.61	26.30
Flavone	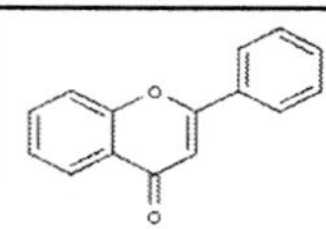	-	3.56	1.37	179.2	64.20	25.45	26.30
Quercetin	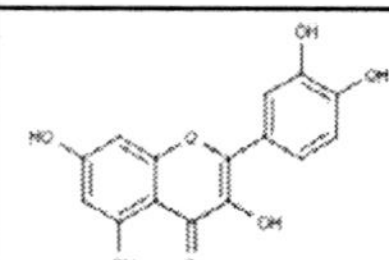	-	1.48	1.11	167.9	73.31	29.06	127.45
Rutin	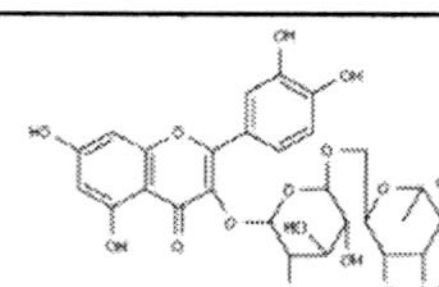	4.30	-2.02	0.36	333.5	138.19	54.78	265.52

Table 4. (Continued)

Drug	Structure	pKa	LogP	$Logk_{BMC}$	V_M (cm^3)	R_M (cm^3)	Pol ($\cdot 10^{24}$) (cm^3)	PSA ($Å^2$)
Vicenin-2	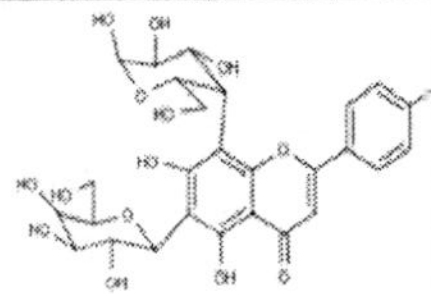	-	-4.31	-0.22	338.1	137.56	54.53	267.29
Vitexin	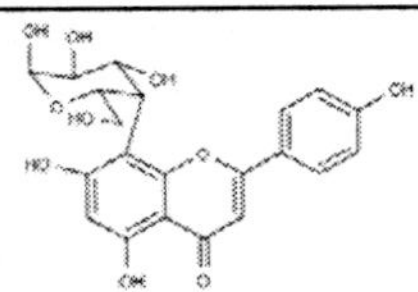	-	0.32	0.67	256.3	103.70	41.11	177.14

As can be observed in Figure 4A, the binding of β-blockers, phenothiazines and antihistamines to HSA (PB_{HSA}) and AGP (PB_{AGP}) is directly correlated with the hydrophobic parameters (logP and $logk_{BMC}$) and inversely correlated with the electronic parameter PSA, which means that hydrophobic and weak polar basic drugs interact mainly with HSA and AGP. However, the binding to lipoproteins and/or globulins ($PB_{LIPO+GLOB}$) seems to be directly partially correlated with PSA and inversely correlated with hydrophobic parameters (logP and $logk_{BMC}$). Steric (V_M, R_M, Pol) parameters showed no correlation with protein binding

As regards acidic drugs such as barbiturates, benzodiazepines and polyphenolic compounds, their binding to HSA is also directly correlated with the hydrophobic parameters and inversely correlated with the electronic parameter PSA but their binding to AGP is directly correlated with the steric parameters (V_M, R_M, Pol).

CONCLUSION

Protein binding of drugs is a crucial factor affecting their pharmacokinetics, pharmacodynamics and toxicity since it determines the free drug concentration. Several methodologies have been reported in the literature to evaluate the binding of drugs to plasma proteins. Among them, capillary electrophoresis in the frontal analysis modality seems to be the preferred choice due to its simplicity, speed, low protein comsumption. Moreover, the injection of sample in the short-end of the capillary reduces the analysis time. The mathematical data treatment proposed to determine the affinity parameters to proteins (number of binding sites and affinity constants) reduces errors in the estimation. Consequently, both proposals have significantly improved the frontal analysis methodology performance to evaluate the drug-protein interaction.

The results showed in this chapter reveals that HSA has not only affinity towards neutral and acid drugs such as barbiturates, benzodiazepines and polyphenolic compounds, but also interacts with hydrophobic basic compounds such as β-blockers, phenothiazines and antihistamines. The interaction with HSA of compounds studied is determined by their hydrophobicity (direct relationship) and the electronic parameters (indirect relationship) of the compounds.

The binding to AGP depends on the compounds; for basic compounds: β-blockers, phenothiazines, antihistamines the interaction with AGP is determined by the hydrophobicity (direct relationship) and indirectly by the polar surface area (indirect relationship). For neutral or acid compounds, i.e. barbiturates, benzodiazepines and polyphenols, this interaction only depends on the steric parameters (direct relationship).

ACKNOWLEDGMENTS

The authors acknowledge the Spanish Ministry of Science and Innovation (MICINN), (Project SAF2008-00859) for the financial support. L. Escuder-Gilabert is grateful to the Generalitat Valenciana for the grant (APOSTD/2007/062). S. Sagrado acknowledges the Spanish Ministry of Science and Technology (MCYT) (project HS2008-0002) for the financial support.

REFERENCES

[1] Su, T. J., Lu, J. R., Cui, Z. F. & Thomas, R. K. (2000). Fouling of ceramic membranas by albumins Ander dynamic filtration conditions. *J. Membr. Sci.*, *173*, 167-178.

[2] González-Alonso, I. & Sánchez-Navarro, A. (1998). *Unión a proteínas. En Biofarmacia y Farmacocinética* (J., Doménech Berrozpe, J. & Martínez Lanao, J. M. Plá Delfina, Editores)". Volumen 2. Síntesis, Madrid.

[3] Sugio, S., Kashima, A., Mochizuki, S., Noda, M. & Kobayashi, K. (1999). Cristal structure of human serum albumin at 2.5 Å resolution. *Protein Engineering*, *12*, 439-446.

[4] McDonnell, P. A., Caldwell, G. W. & Masucci, J. A. (1998). Using capillary electrophoresis/frontal analysis to screen drugs interacting with human serum proteins. *Electrophoresis*, *19*, 448-454.

[5] Busch, M. H. A., Kraak, K. C. & Poppe, H. (1997)[a]. Principles and limitations of methods available for the determination of binding constants with affinity capillary electrophoresis. *J. Chromatography A*, *777*, 329-353.

[6] Evans, W. E., Schentag, J. J., Jusko, W. J. & Relling, M. V. (1992).

Applied Pharmacokinetics. Principles of therapeutic drug monitoring. Edwards Brothers, Vancouver.
[7] Müller, W. E. & Wollert, U. (1979). Human serum albumin as a silent receptor for drugs and endogenous substances. *Pharmacology, 19*, 59-67.
[8] Sjöholm, I. (1986). *In Drug-Protein binding*, (Reidenberg M.M., Erill S., editors). Praeger Publishers, New York.
[9] Müller, W. E., Fehske, K. J. & Schläfer, S. A. C. (1986). *In Drug-Protein binding*, (Reidenberg M.M., Erill S., editors). Praeger Publishers, New York.
[10] Sudlow, G., Birkett, D. J. & Wade, D. N. (1975). The characterization of two specific drug binding sites on human serum albumin. *Mol. pharmacol. 11*, 824-832.
[11] Fehske, K. J., Müller, W. E. & Wollert, U. (1981). The location of drug binding sites in human serum albumin. *Biochem. Pharmacol, 30*, 687-692.
[12] Ghuman, J., Zunszain, P. A., Petitpas, I., Bhattacharya, A. A., Otagiri, M. & Curry, S. (2005). Structural basis of the drug-binding specificity of human serum albumin. *J. mol. biology, 353*, 38-52.
[13] Hage, D. S. & Sengupta, A. (1998). Studies of protein binding to nonpolar solutes by using zonal elution and High- Performance Affinity Chromatography: Interactions of cis- and trans-clomiphene with human serum albumin in the presence of β-cyclodextrin. *Anal. Chem., 70*, 4602-4609.
[14] Hage, D. S. & Sengupta, A. (1999). Characterization of the binding of digitoxin and acetyldigitoxin to human serum albumin by high-performance affinity chromatography. *J. Chromatogr. B, 724*, 91-100.
[15] Sengupta, A. & Hage, D. S. (1999). Characterization of minor site probes for human serum albumin by high-performance affinity chromatography. *Anal. Chem., 71*, 3821-3827.
[16] Haginaka, J. (2000). Enantiomer separation of drugs by capillary electrophoresis using proteins as chiral selectors. *J. Chromatogr. A, 875*, 235-254.
[17] Zsila, F. & Iwao, Y. (2007). The drug binding site of α_1-acid glycoprotein: Insight from induced circular dichroism and electronic absorption spectra. *Biochim. Biophys. Acta, 1770*, 797-809.
[18] Kremer, J. M. H., Wilting, K. & Janssen, L. H. M. (1988). Drug binding to human alfa-1-acid glycoprotein in health and disease. *Pharmacol. reviews, 40*, 1-47.

[19] Doménech Berrozpe, J., Martínez Lanao, J. & Plá Delfina, J. M. (1998). *Biofarmacia y farmacocinética. (Volumen 2: Biofarmacia)*. Síntesis, Madrid.

[20] Jiménez-Torres, N. V., Casabó-Alós, V. G. & Sancho Chust, V. (1997). *Manual de procedimientos para Farmacocinética Clínica.* 1st ed., AFAHPE, Valencia.

[21] Tanaka, Y. & Terabe, S. (2001). Recent advances in enantiomer separations by affinity capillary electrophoresis using proteins and peptides. *J. Biochem. Biophys. Methods*, *48*, 103-116.

[22] Kwong, T. C. (1985). Free drug measurements: methodology and clinical significance. *Clin. Chim. Acta*, *151*, 193-216.

[23] Kurz, H. (1986). *Methodological problems in drug binding studies-In Drug Protein Binding*. Praeger Publishers, New York.

[24] Ascoli, G., Bertucci, C. & Salvadori, P. (1995). Stereospecific and competitive binding of drugs to human serum albumin: a difference circular dichroism approach. *J. Pharm. Sci.*, *84*, 737-741.

[25] Abadín, J. A. & Durán, J. A. (1989). *Unión de los fármacos a las proteínas y su trascendencia clínica*. Abbot científica, Sevilla.

[26] Tanaka, M., Asahi, Y., Masuda, S. & Ota, T. (1991). Binding position of azathioprine with bovine serum albumin determined by measuring nuclear magnetic resonance relaxation time. *Chem.Pharm. Bullet*, *39*, 2771-2774.

[27] Alexander, K. A. & Phelps, W. C. (1996). A fluorescence anisotropy study of DNA binding by HPV-11 E2C protein: A hierarchy of E2-binding sites. *Biochemistry*, *35*, 9864-9872.

[28] Bertucci, C. & Cimitan, S. (2003). Rapid screening of small ligand affinity to human serum albumin by an optical biosensor. *J. Pharm. Biomed. Anal.*, *32*, 707-714.

[29] Heegard, N. H. H., Nilsson, S. & Guzman, N. A. (1998). Affinity capillary electrophoresis: important application areas and some recent developments. *J. Chromatogr. B*, *715*, 29-54.

[30] Hage, D. S. (2002). High-performance affinity chromatography: a powerful tool for studying serum protein binding. *J. Chromatogr. B*, *768*, 3-30.

[31] Loun, B. & Hage, D. S. (1996). Chiral separation mechanisms in protein-based HPLC columns. 2. Kinetic studies of (R)- and (S)-warfarin binding to immobilized human serum albumin. *Anal. Chem.*, *68*, 1218-1225.

[32] Noctor, T. A. G., Díaz-Perez, M. J. & Wainer, I. W. (1993). Use of a

human serum albumin-based stationary phase for high-performance liquid chromatography as a tool for the rapad determination of drug-plasma protein bindinb. *J. Pharm. Sci.*, *82*, 675- 676.

[33] Singh, S. S. & Mehta, J. (2006). Measurement of drug-protein binding by immobilized human serum albumin-HPLC and comparison with ultrafiltration. *J. Chromatogr. B*, *834*, 108-116.

[34] Loun, B. & Hage, D. S. (1992). Characterization of thyroxine-albumin binding using high-performance affinity chromatography I. Interactions at the warfarin and indole sites of albumin. *J. Chromatogr.: Biomed. Applic*, *579*, 225-235.

[35] Qiao, M., Guo, X. & Li, F. (2002). Chemiluminiscence detection coupled to high-performance frontal análisis for the determination of unbound concentrations of drugs in protein binding equilibrium. *J. Chromatogr. A*, *952*, 131-138.

[36] Zhang, W., Zhang, L., Ping, G., Zhang, Y. & Kettrup, A. (2002). Study on the multiple sites binding of human serum albumin and porphyrin by affinity capillary electrophoresis. *J. Chromatogr. B*, *768*, 211-214.

[37] Tanaka, Y. & Terabe, S. (2002). Estimation of binding constants by capillary electrophoresis. *J. Chromatogr. B*, *768*, 81-92.

[38] Busch, M. H. A., Carels, L. B., Boelens, H. F. M., Kraak, K. C. & Poppe, H. (1997)[b]. Comparison of five methods for the study of drug-protein binding in affinity capillary electrophoresis. *J. Chromatogr. A*, *777*, 311-328.

[39] He, X., Ding, Y., Li, D. & Lin, B. (2004). Recent advances in the study of biomolecular interactions by capillary electrophoresis. *Electrophoresis*, *25*, 697-711.

[40] Martínez-Plà, J. J., Martínez-Gómez, M. A., Martín-Biosca, Y., Sagrado, S., Villanueva-Camañas, R. M. & Medina-Hernández, M. J. (2004). High-throughput capillary electrophoresis frontal analysis method for the study of drug interactions with human serum albumin at near-physiological conditions. *Electrophoresis*, *25*, 3176-3185.

[41] Østergaard, J., Schou, C., Larsen, C. & Heegaard, N. H. H. (2002). Evaluation of capillary electrophoresis-frontal análisis for the study of low molecular weight drug-human serum albumin interactions. *Electrophoresis*, *23*, 2842-2853.

[42] Le Saux, T., Varenne, A., Perreau, F., Siret, L., Duteil, S., Duhau, L. & Gareil, P. (2006). Determination of the binding parameters for antithrombin-heparin fragment systems by affinity and frontal análisis continuous capillary electrophoresis. *J. Chromatogr. A*, *1132*, 289-296.

[43] Zhang, B., Fung, Y., Lau, K. & Lin, B. (1999). Bilirubin-human serum albumin interaction monitored by capillary zone electrophoresis. *Biomed. Chromatogr*, *13*, 267-271.

[44] Ding, Y., Lin, B. & Huie, C. W. (2000). Thermodynamics of porphyrin binding to human serum albumin using affinity capillary electrophoresis. *Cromatographia*, *52*, 367-370.

[45] Amini, A. & Werterlund, D. (1998). Evaluation of Association constants between drug enantiomers and Human α_1-acid glycoprotein by applying a patial-filling technique in affinity capillary electrophoresis. *Anal. Chem.*, *70*, 1425-1430.

[46] Sabah, S. & Scriba, G. K. E. (1999). pH-dependent reversal of the chiral recognition of tripeptide enantiomers by carboxymethyl-β-cyclodextrin. *J. Chromatogr. A*, *833*, 261-266.

[47] Ding, Y., Zhu, X. & Lin, B. (1999). Study of interaction between drug enantiomers and serum albumin by capillary electrophoresis. *Electrophoresis*, *20*, 1890-1894.

[48] Ishihama, Y., Miwa, T. & Asakawa, N. (2002). Drug-plasma protein binding assay by electrokinetic chromatography-frontal analysis. *Electrophoresis*, *23*, 951-955.

[49] McDonnell, P. A. & Caldwell, G. W. (1997). *Recent Advances in Analytical Techniques*, Gordon and Breach Science Publisher, Amsterdam.

[50] Hardman, J. G., Limbird, L. E., Molinoff, P. B., Ruddon, R. W. & Goodman Gilman, A. (1996). *Goodman & Gilman. Las bases farmacológicas de la Terapéutica*. 9th ed., McGraw-Hill Interamericana, Ciudad de México, México.

[51] Martínez-Gómez, M. A., Villanueva-Camañas, R. M., Sagrado, S. & Medina-Hernández, M. J. (2006). Characterization of basic drug-human serum proteins interactions by capillary electrophoresis. *Electrophoresis*, *27*, 3410-3419.

[52] Martínez-Gómez, M. A., Carril-Avilés, M. M., Villanueva-Camañas, R. M., Sagrado S. & Medina-Hernández, M. J. (2007). Characterization of antihistamines-human serum proteins interactions by capillary electrophoresis. *J. Chromatogr. A*, *1147*, 261-269.

[53] Diniz, A., Escuder-Gilabert, L., Lopes, N. P., Villanueva-Camañas, R. M., Sagrado, S. & Medina-Hernández, M. J. (2008). Characterization of interactions between polyphenolic compounds and human serum proteins interactions by capillary electrophoresis. *Anal. Bioanal. Chem.*, *391*, 625-632.

In: Biomedical Chromatography
Editors: John T. Elwood, pp. 69-103
ISBN: 978-1-60741-291-5

Chapter 3

STATISTICAL APPROACHES FOR BIOEQUIVALENCE OF HIGHLY VARIABLE DRUGS AND DRUG PRODUCTS

***Alionka Citlali P. Angeles-Moreno*[1], *Gabriel Marcelín- Jiménez*[1*] *and Salvador Zamora-Muñoz*[2]**
[1]Clinical Pharmacology Research Service, Hospital, General de México, Mexico City, Mexico
[2]IIMAS, Universidad Nacional Autónoma de México (UNAM), Mexico City, Mexico

GENERAL BACKGROUND

Bioequivalence studies play a major role in the development of new drugs and in the marketing of generic formulations; such trials also contribute in to access to low-cost and effective medicines in developing countries. At present, with the loss of patents of novel molecules, the difficulty in designing interchangeability trials has increased.

* Corresponding author: Servicio de Investigación de Farmacología Clínica (405-E), Hospital General de México, Dr. Balmis No. 148, Col. Doctores, 06726 México, D.F., México. E-mail: gabmarcelin@hotmail.com

Several factors contribute to the complexity of bioequivalence studies. Some new drugs are more potent, and their concentrations in biological fluids are lower each time. In addition to this, the task of controlling the content uniformity of the formulations of these drugs is more demanding

Finally, due to the global politics of generic usage, it has become evident that there is minimal information concerning metabolic background in different populations, which has comprised another unexpected source of variation.

The aim of a bioequivalence study is to demonstrate in a controlled manner that there are no significant differences between an innovator and a generic formulation regarding the amount and speed of drug absorption. We can infer clinical effectiveness and safety when the difference in bioavailability of both generic and innovator is <20%. When logarithm data of measurements are related, this gives bioequivalence limits of 0.8– 1.25.

Currently, the evaluation of bioequivalence for products exhibiting High Intra-individual variability (highly variable drugs [HIV]) or high variation in the formulation (highly variable drug products [HVP]) has become a serious problem. This is due to that the shape of normal distribution of pharmacokinetic data widens. Consequently, there is an increase in Confidence intervals (CI), which renders detection of possible differences or similarities among products difficult.

Therefore, on reducing the range of the CI, it is necessary to increase the sample size, which translates into a substantial increase in the study cost and in the risk of drop-outs.

Blume et al. [1] proposed the multiple-dose design for testing bioequivalence during HVD and HVP steady state. This design has been questioned when products with low quality and high variation in their formulation are studied, because the design can mask variations in production, delivering low-quality into the market [2].

Another proposed method for approaching HVD and HVP bioequivalence studies has been increasing the bioequivalence limits (extending the criteria to 0.75–1.33 instead of 0.80–1.25) [3]. However, this is only useful for products with high therapeutic margin and safety. When such a conditions are not satisfied, this criterion may place therapeutic effectiveness at risk, and the approach problem persists.

Some regulatory agencies have published guidelines that describe the general procedures for evaluating bioequivalence. One example is the U.S. Food and Drug Administration (FDA), which in 2001 published "Statistical Approaches to Establishing Bioequivalence" [4]. In these guidelines, two very important concepts were introduced: Population- and Individual-Bioequivalence. These two approaches are employed depending on the trials' ultimate objectives. Individual bioequivalence has been the most frequently used tool for approaching HVD and HVP.

OBJECTIVE

The aim of the present chapter is to show that in the traditional 2x2 bioequivalence design, residual error is comprises different, not properly isolated sources of variation (subject variation, analytical variations, formulation variations, and random variations), and in turn, a huge sample size for obtaining statistical power.

The individual bioequivalence model is explained mid-chapter. ANOVA is performed in a very clear fashion, exhibiting each of the isolated sources of variation. Sample size and power in the 2x3 and 2x4 designs are compared with results of traditional design of average bioequivalence.

Finally, we discuss new trends for approaching individual bioequivalence.

HIGHLY VARIABLE DRUGS AND DRUG PRODUCTS

In bioequivalence studies, three main sources of variation can be detected:

a) Intra-subject, b) Between-subject, and c) Within-formulation variations. Each is described as follows:

Intra-Subject Variation

Intra-subject variation is closely related with HVD. A drug is considered highly variable when Intra-subject variation is ≥30% C.V. of pharmacokinetic parameters (Maximum concentration [Cmax] and/or Area under the curve [AUC]) [3].

This means that the variations of the previously defined measurements between each dose are very different for each subject.

This concept is better understood in Figure 1. As can be observed, Cmax and AUC are quite different in the same subject, with the same dose of different formulations, at different dosing time.

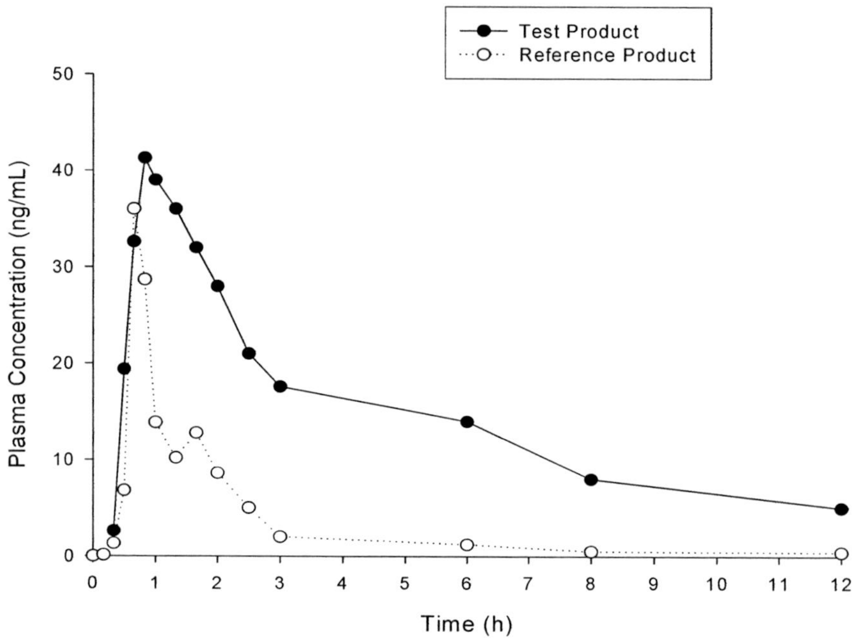

Figure 1. Pharmacokinetic profile of one volunteer who takes Clopidogrel Test formulation during first period, and Reference formulation during second one (Courtesy of Hospital General de Mexico).

Between-Subject Variation

This is defined as the variation in measurements among subjects who take the same product, at the same dosage, at the same time. This variation is attributable to a high degree of polymorphism in the enzymes responsible for biotransformation.

One example of this is the family of proton-pump inhibitors (omeprazole, pantoprazole, lansoprazole). These drugs undergo extensive first-pass metabolism by means of isoform 2C19 of CYP450, which presents five variants.

An example of this behavior is depicted in Figure 2. Two volunteers were dosed with 150 mg of Clopidogrel in two different periods. One volunteer showed a Cmax for both formulations of ca. 2000 ng/mL, while the second exhibited a Cmax of 16,000 ng/mL for both formulations. It is evident that there are sub-populations in the sample of the trial.

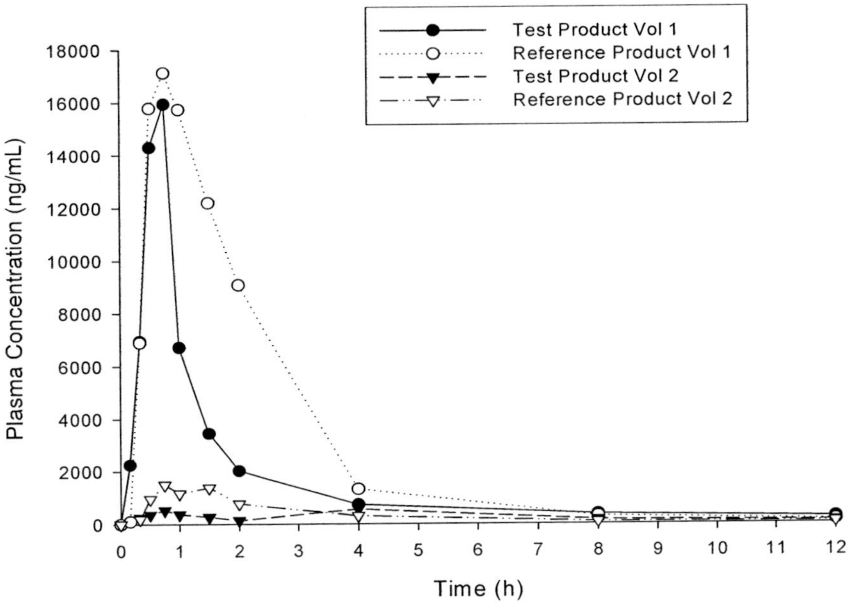

Figure 2. Pharmacokinetic profiles of Clopidogrel from two volunteers, who take in the first period the Test product, and in the second period the Reference product (Courtesy of Hospital General de Mexico).

Within-Formulation Variation

This variation is observed when a product of the same batch is administered at the same time to several subjects (e.g., tablet-to-tablet variation). The measurements of some subjects are different compared to the average response. This could be attributable to the drug itself in a poor-quality product [3, 10, 12].

In Figure 3, it is possible to note that the population exhibits a homogeneous pharmacokinetic profile; however, three subjects (1, 5, and 8) showed a different profile compared with the average. This could be attributed to tablet-to-tablet variations, which would refer to a low-quality manufacturing process.

Another example of Within-formulation variation is depicted in Figure 4. Four groups (24 subjects per group) from Mexican population were administered a single dose of the same product from different batches of Loratadine tablets.

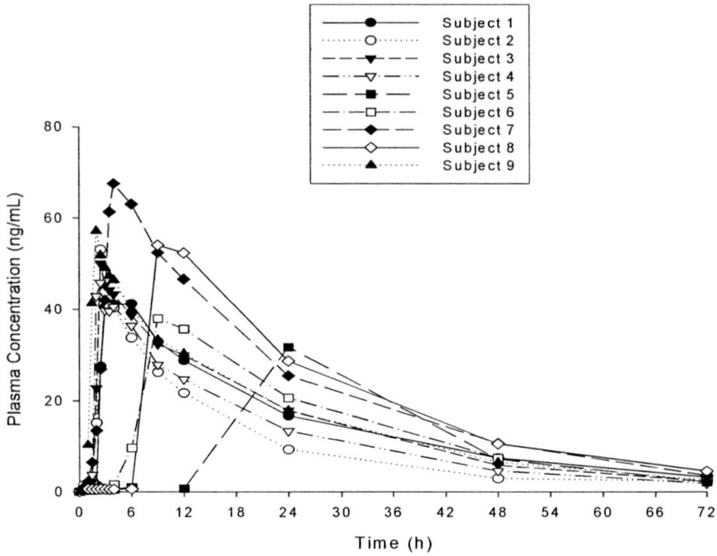

Figure 3. Pharmacokinetic profiles of magnesium valproate from nine volunteers, who take the same product (tablets with the same concentration), at the same time (Courtesy of Hospital General de Mexico).

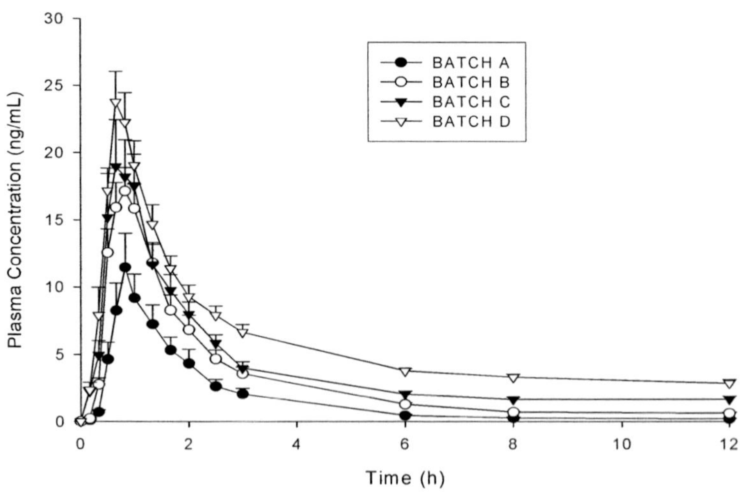

Figure 4. Average Pharmacokinetic profile from 4 different groups administered with an oral single dose of Loratadine tablets from 4 different batches of the same product (Courtesy of Hospital General de Mexico).

When bioequivalence is assayed among different batches, it can be concluded that only batches B and C are bioequivalent. When batches B and D are compared, the result is no-bioequivalence. The same result, non-bioequivalence, is obtained when batches A and C are tested. The most extreme result is observed when batches A and D are compared.

CONSIDERATIONS FOR BIOEQUIVALENCE TRIAL DESIGN

When a bioequivalence trial is to be conducted, it is extremely important to consider several aspects that exert a direct impact on statistical power, therefore, on the strength of the conclusion. Such aspects are described here:

1. **Population.** This is defined as the sample of humans in whom the evaluation will be performed. The population must be sufficiently homogenous to avoid possible confusion between 1) biological variability and 2) formulation-related variations. The demographic variables that must be under control are weight, height, age, and in some cases, gender. The healthy condition of volunteers is an additional characteristic that must be evaluated in order to isolate variations that could interfere with the experimental model.
   If bioequivalence is assayed in non-healthy persons, other considerations must be taken into account: possible interference of concomitant medication, degree of disease, etc.
   - 1.a. **Reported biological variability.** This factor is not always analyzed, because there are a greater number of drugs than do not possess sufficient information on their variability in different populations. Thus, it is recommendable for these drugs to perform pilot studies to define biological variability in the target population. This point is fundamental, not only for sample size calculation, but also for the type of design that will be conducted (average bioequivalence or individual bioequivalence)
   - 1.b. **Metabolism.** This factor is important because if there is evidence indicating that this is a drug with a probable carryover effect, this is due to performing experimental designs that allow for isolation and to quantifying this effect. One example of these designs is the Balaam design [5], which utilizes four sequences and two periods, or 2x3 or 2x4 crossover individual bioequivalence designs.

2. **Dose.** The dosage amount and schedule must be considered. Dosing schedules can be a single- or multiple-dose; the latter is accepted as diminishing biological variability, and when modified, delivering formulations are tested. The multi-dose schedule may increase the risk of side events in volunteers during the trial.

3. **Sampling schedule.** An adequate sampling-time design is an important process that allows for better description of the pharmacokinetic profile, which is associated with more precise calculations of related measurements. During the sampling-time schedule design, it is very important to establish the washout period of the drug. Thus, errors associated with drug carryover are avoided. According to the Mexican regulatory framework [6], a washout period must be of at least 7 half-lives of the parent drug or of the main metabolites (which will be the largest in size).

4. **Anticoagulants and non-planned co-medication.** It is necessary to analyze interferences of the drug with the possible anticoagulants used during sampling in order to select the better one.

In addition to anticoagulants, it is sometimes necessary to employ a certain type of medication when a volunteer refers a slight inconvenience (headache, stomachache, toothache, etc.). In these cases, it is necessary to establish the possible interferences of these medications.

5. **The number of formulations that will be evaluated (two or more).** This factor is very critical, because the number of periods and sequences associated will be dependent on the number of products to be evaluated.
   This item should also be considered when a balanced or non-balanced design will be used.
   A balanced design is that in which:
   - Each formulation is administered to each subject once.
   - Each formulation is present the same number of times in every period.
   - The number of volunteers assigned to sequence 1 is the same as that assigned to the mirror sequence.

6. **Length of treatment.** With drugs having a very long half-life ($t_{1/2}$ >24 h), it is necessary to consider the length of the sampling time and the washout period. It is more advisable to consider parallel designs or crossover truncated designs.
   Very long treatment periods may increase the probability of dropouts. This should be considered from the beginning of the trial, because it means an increase in sample size.

7. **Running-protocol considerations.** There are several activity and condition types that must be standardized during each period of dosing. These conditions are as follows a) Physical activity of volunteers; b) sleep time; c) fasting period; d) diet (meals and beverages), and in some cases, e) temperature collection of samples and f) type of light during plasma separation and drug quantification (depending on the light susceptibility of the molecule).

Analysis of Variance (ANOVA) for Average Bioequivalence Studies

The aim of average bioequivalence is to demonstrate the interchangeability of two formulations, considering the general variability of one population.

If we analyze the design for the evaluation of average bioequivalence in detail, it can be observed that [7]:

$$Y_{ijk} = \mu + S_{ik} + P_j + F_{jk} + C_{(j-1k)} + e_{ijk} \qquad \text{(Eq. 1)}$$

where:

Y_{ijk} = is the log response of the measurement ($ABC_{0\text{-}t}$, $ABC_{0\text{-}inf}$, or Cmax) of subject i^{th} in k^{th} sequence of jth period.

μ = the overall mean.

S_{ik} = the randomized effect of subject i^{th} in k^{th} sequence.

P_j = the effect of jth period.

F_{jk} = the effect of the formulation in k^{th} sequence, which has been administered in j^{th} period.

$C_{(j-1k)}$ = the first-order carryover effect of the formulation in k^{th} sequence, which is administered in (j-1) period.

e_{ijk}= the residual error.

As can be noted, carryover effect is considered in the model; however, in 2x2 designs, this factor is assumed as zero. Such an assumption is based on that a sufficiently lengthy washout period is in place between dosing periods.

In this model, there are several sources of variation that must be evaluated to define whether the experimental design fulfills the following assumptions:

The period effect. This effect could be defined as changes in the activities run during each period, or changes in the average response of the organism from one period to the next.
It is important to verify that a significant difference between both periods does not exist.

The product effect. This is another source of variation analyzed during ANOVA. The presence of such an effect does not comprise a condition for design rejection, because it is necessary to confirm the magnitude of the difference through hypothesis tests.

The sequence effect. This effect is a reflection of a well-done randomization. It is analyzed to evaluate whether volunteers assigned to one sequences do not differ significantly from volunteers who took the mirror sequence (TR or RT). When in a single-dose design the sequence effect exists, this effect must be studied and dissected very well, because frequently the sequence effect is confused with the carryover effect.

The aim of ANOVA is to confirm that the experimental design was performed correctly.

In agreement with the 2x2 crossover design, there are three different factors that are important to identify and evaluate. Typical ANOVA data are shown in Table 1.

It can be observed that each source of variation (factor) has two levels. For example, the Product factor has two levels: test product, and reference product. The period factor includes the first and the second period. And the sequence factor has the TR sequence and the mirror (RT) sequence.

Each level has several observations that correspond with the measurement (Cmax or AUC). The amount of observations depends on the number of subjects assigned to each previously discussed factor.

Table 1. Scheme for data presentation for typical Analysis of Variance (ANOVA)

FACTOR	LEVEL	VARIABLE
PRODUCT	A	AUC (1)
		AUC (2)
		AUC (3)
		AUC (4)
		AUC (5)
	B	AUC (1)
		AUC (2)
		AUC (3)
		AUC (4)
		AUC (5)
PERIOD	1	AUC (1)
		AUC (2)
		AUC (3)
		AUC (4)
		AUC (5)
	2	AUC (1)
		AUC (2)
		AUC (3)
		AUC (4)
		AUC (5)

FACTOR	LEVEL	VARIABLE
SEQUENCE	AB	AUC (1)
		AUC (2)
		AUC (3)
		AUC (4)
		AUC (5)
	BA	AUC (1)
		AUC (2)
		AUC (3)
		AUC (4)
		AUC (5)

ANOVA is important for identifying the total variation present in the whole population, which is composed by Between- and Within-subject variations.

Variation is defined as the distance between each observation and the average of the corresponding measurement.

Total variation can be expressed as:

$$SS_{total} = SS_{between} + SS_{within} \quad \text{(Eq. 2)}$$

where:

SS: Sum of Squares.

Table 2 depicts Within- and Between-subject variation. The vertical arrow indicated Between-subject variation, which is composed of the carryover effect and the residual error (Inter), while the horizontal arrow demonstrates the Within-subject variation, which is associated with variations present by each volunteer in each period with each product.

The mathematical expression of this is:

$$SS_{between} = SS_{carry} + SS_{Inter} \qquad \text{(Eq. 3)}$$

SS between subjects considers Between-subject carryover effect and residuals.

The carryover effect can be expressed as:

$$SS_{Carry} = \frac{2n_1 n_2}{n_1 n_2}\left\{(\bar{Y}_{\cdot 12} + \bar{Y}_{\cdot 22}) - (\bar{Y}_{\cdot 11} + \bar{Y}_{\cdot 21})\right\}^2 \qquad \text{(Eq. 4)}$$

Bearing in mind that in a 2x2 crossover design there are two periods and two sequences:

TR-----sequence 1
RT-----sequence 2

Table 2. Graph showing Between- and Within-subject variations

PERIOD		1	2
	PRODUCT	T	R
SEQUENCE	SUBJECTS	1	1
		2	2
		5	5
		6	6
		9	9
		11	11
		13	13
PRODUCT		R	T
SEQUENCE	SUBJECTS	3	3
		4	4
		7	7
		8	8
		10	10
		12	12
		14	14

Thus, in Eq. 4, $Y_{\cdot 12}$ represents the average of the measurement (Cmax, AUC) for product R during the first period in Sequence 2, and $Y_{\cdot 22}$ represents the average of the same measurement for product T during the second period in the same Sequence 2. The term $Y_{\cdot 11}$ represents the average of the measurements for product T during the first period in Sequence 1, and finally, $Y_{\cdot 21}$ represents the average of the measurement for product R during the second period in the same Sequence 1.

Likewise, Between-subject residuals can be expressed as:

$$SS_{inter} = \sum_{k=1}^{2}\sum_{i=1}^{n_k}\frac{Y^2_{i.k}}{2} - \sum_{K=1}^{2}\frac{Y^2_{..K}}{2n_k} \qquad \text{(Eq. 5)}$$

On the other hand, Within-subject SS is defined as:

$$SS_{within} = SS_{drug} + SS_{period} + SS_{intra} \qquad \text{(Eq. 6)}$$

whereas Within variability is made up of the following components:

$$SS_{drug} = \frac{2n_1n_2}{n_1+n_2}\left\{\frac{1}{2}\left[(\bar{Y}_{\cdot 21}-\bar{Y}_{\cdot 11})-(\bar{Y}_{\cdot 22}-\bar{Y}_{\cdot 12})\right]\right\}^2 \qquad \text{(Eq. 7)}$$

$$SS_{period} = \frac{2n_1n_2}{n_1+n_2}\left\{\left[(\bar{Y}_{\cdot 21}-\bar{Y}_{\cdot 11})-(\bar{Y}_{\cdot 12}-\bar{Y}_{\cdot 22})\right]\right\}^2 \qquad \text{(Eq. 8)}$$

$$SS_{intra} = \sum_{K=1}^{2}\sum_{j=1}^{2}\sum_{i=1}^{n_k}Y^2_{ijk} - \sum_{k=1}^{2}\sum_{i=1}^{n_k}\frac{Y^2_{i.k}}{2} - \sum_{k=1}^{2}\sum_{j=1}^{2}\frac{Y^2_{.jk}}{n_k} + \sum_{k=1}^{2}\frac{Y^2_{..k}}{2n_k} \qquad \text{(Eq. 9)}$$

In order to understand ANOVA easily, we will practice it, using bioequivalence study data of Clopidogrel.

According to that mentioned in the present chapter, it is important to consider the biological variability reported in the references. Biological variability of ca, 20% has been described for this drug [8].

However, other authors mention a higher variability (>30%); thus, we decided to consider an *n* of 27 subjects for every sequence in this study.

Table 3 presents data from each AUC period of a bioequivalence study developed in a Mexican population using a 150-mg dose of Clopidogrel.

To conduct ANOVA, the following SS sums must be performed:

$$SS_{total} = SS_{between} + SS_{within} \quad \text{(Eq. 2)}$$

$$SS_{between} = SS_{carry} + SS_{Inter} \quad \text{(Eq. 3)}$$

$$SS_{within} = SS_{drug} + SS_{period} + SS_{intra} \quad \text{(Eq. 6)}$$

Table 3. AUC (ng*h/ml) for test and reference products of 75 mg Clopidogrel Tablets in a Mexican population (Dose, 150 mg)

Period	I	II	Period	I	II
Formulation	T	R	Formulation	R	T
Subject			Subject		
1	3121.8922	1763.1141	2	19424.5158	37447.6481
4	5870.3354	7421.8748	3	3130.4449	3976.8757
8	3807.5508	5841.624	5	4281.0767	5095.3792
9	8777.885	4474.0528	6	23411.2375	5407.433
12	1150.5636	4304.0521	7	4462.185	1051.6397
13	9693.3213	12027.6552	10	4082.9151	3084.1172
14	13091.0223	4874.3464	11	2784.5772	1826.665
16	9396.3471	8951.0671	15	14657.9096	8275.8502
22	3331.7046	4221.682	17	5152.2724	5005.0693
23	2043.3594	1264.4892	18	1599.5919	1899.6859
24	1108.4635	666.3351	19	5277.1003	7279.7476
26	3930.9644	4213.6787	20	2859.3206	3167.896
27	5601.1001	2734.2607	21	5113.715	5586.0927
28	2076.6798	5457.4522	25	4437.8766	4933.0243
30	26844.8136	1782.4973	29	10319.9419	3877.2869
32	9975.2708	13501.4172	31	4677.1492	3066.1502
33	2104.8001	3575.8038	34	1495.9806	1820.6917
36	1831.6488	4059.5406	35	5829.9748	3876.8342
37	3582.5536	8176.4659	39	13218.0943	10898.9702
38	3838.7369	8806.4188	42	4736.0973	5925.6667
40	10279.0579	17587.6641	44	9093.1888	7850.1596
41	6112.9797	11173.0186	49	7218.1126	8747.6516
43	3383.591	2756.4446	50	7890.6908	9005.6907
45	5300.2424	4841.5516	51	3654.5755	3299.8735
46	9103.2458	8486.9511	52	4094.4953	4941.9954
47	10234.4432	4889.7266	53	17068.8405	18490.7252
48	8399.524	17983.3899	54	12026.7689	13930.9337

Table 4 shows the different sums required to carry out ANOVA. These sums are grouped by each factor. First, the sums for each period appear, followed by the sums for each sequence.

Table 4. Example Of Sum Of Periods And Sequences Necessary For Conducting ANOVA

	Period	I	II		Period	I	II
	Formulation	T	R		Formulation	T	R
Sequence	Subject	SUM	SUM	SEQUENCE	Subject	SUM	SUM
	1				2		
	4				3		
	8				5		
	9				6		
	12				7		
	13				10		
	14				11		
	16				15		
	22				17		
	23				18		
	24				19		
	26				20		
	27				21		
1 (TR)	28	173992.097	175836.575	2 (RT)	25	201998.649	189769.754
	30				29		
	32				31		
	33				34		
	36				35		
	37				39		
	38				42		
	40				44		
	41				49		
	43				50		
	45				51		
	46				52		
	47				53		
	48				54		

Table 5. Condensed table for every source of variation

SUM FORMULATION	T	R
	363761.8508	377835.2236
SUM PERIOD	I	II
	375990.7464	365606.328
SUM SEQUENCE	TR	RT
	349828.672	391768.403

$$Error_{intra} = SS_{Xtotal} - (SS_{Seq} + SS_{drug} + SS_{Period} + SS_{Subject(seq)}) \quad \text{(Eq. 9)}$$

Table 5 depicted in a condensed manner the sums for every source of variation: sums for test and reference products for each period and each sequence.

The following sums must also be performed:

$$SS_{XTOT} = \sum X_{Total} - \frac{(\sum X_{Total})^2}{N} \quad \text{(Eq. 10)}$$

$$SS_{X^2TOT} = \sum X^2_{Total} - \frac{(\sum X_{Total})^2}{N} \quad \text{(Eq. 11)}$$

Equations 10 and 11 give the following results:

S_{XTOT}	**741597.074**
S_{X^2TOT}	**8591390173**

N	**108**

By substituting each value in each element in the previously noted equations, we obtain:

$$SS_{Seq} = \frac{(\sum Seq(TR))^2 + (\sum Seq(RT))^2}{54} - \frac{(\sum X_{Tot})^2}{N} \quad \text{(Eq. 4a)}$$

$$SS_{Drug} = \frac{(\sum Drug(T))^2 + (\sum Drug(R))^2}{54} - \frac{(\sum X_{Total})^2}{N} \quad \text{(Eq. 7)}$$

$$SS_{Period} = \frac{(\sum PeriodI)^2 + (\sum PeriodII)^2}{54} - \frac{(\sum X_{Total})^2}{N} \quad \text{(Eq. 8)}$$

$$SS_{Subject(Seq)} = \frac{\sum X^2}{2} - \frac{(\sum X_{Total})^2}{N} \quad \text{(Eq. 5)}$$

Equation 5 can also be identified as the Between residual error, and equation 9 can be identified as the Within residual error.

Table 6 data are generated with Table 4 data and replacing Eq. 5 and 9. The next step comprises determination of the degrees of freedom as follows:

Sequence: n-1
Drug: n-1
Period: n-1
Subject(seq) [Residual inter] = $n_1 + n_2 - 2$
Error (Residual intra): $n_1 + n_2 - 2$
X_{TOT}: N-1

For construction of the ANOVA table, we need:

Mean square (MS). This is the ratio of SS divided by the degrees of freedom.

F value. This is the ratio of each MS divided by the value of the MS error

MS and F values are calculated with values in Table 6. The final result is the whole ANOVA table (Table 7).

Finally, different probabilities are obtained with F values on employing the degrees of freedom of the numerator (each of the MS) and the denominator (MS error). Probability values <0.05 imply significant differences in the specific factor.

Table 6. Results of the Sum of squares

$(S_{XTOT})^2/N$	5092279821.84
SS_{XTOT}	3499110351.31
SS_{SEQ}	16286490.92
SS_{DRUG}	1833887.24
SS_{PERIOD}	998482.83
$SS_{SUBJECT(SEQ)}$	2582336818.18
ERROR	897654672.14

Table 7. ANOVA Results

SOURCE	SS	DF	MS	F	P
SS_{SEQ}	16286490.92	1	16286490.92	0.94345582	0.3359
SS_{DRUG}	1833887.24	1	1833887.24	0.10623477	0.7458
SS_{PERIOD}	998482.83	1	998482.8288	0.05784085	0.8109
$SS_{SUBJECT(SEQ)}$	2582336818.18	52	49660323.43	2.87675974	0.0010
ERROR	897654672.14	52	17262589.85		
SS_{XTOT}	3499110351.31	107	32701965.9		

When analyzing the probability of each variation source in Table 7, it can be concluded that:

There is no period effect.
There is no sequence effect.
There is no formulation effect.

This can be translated into:

(a) A well-designed and controlled experiment.
(b) A sufficiently lengthy washout period to avoid the carryover effect.

After a satisfactory ANOVA result, the following step is the evaluation of bioequivalence.

With these results and utilizing the Schuirmann's test and classical 90% confidence interval (CI) (Table 8), it is possible to verify that unfortunately, the power is <80%. Therefore, the conclusion in terms of bioequivalence is not useful, and it is necessary to recalculate the sample size in order to achieve sufficient statistical power.

Sample Size Determination

Considering that the objective of a bioequivalence study is to determine, with sufficient power, the degree to which two treatments are similar. This is possible through the use of two statistical approaches: Building CIs, and hypothesis tests.

Table 8. Bioequivalence results

Dependent	Ratio	CL_90_Lower	CL_90_Upper	21Prob<80.00	21Prob>120.00	Power
AUCall	96.2753	77.1330	115.4175	0.0802	0.0214	0.3993

During hypothesis testing, the researcher can commit two types of errors:

The Type I error: This is characterized by rejection of the null hypothesis, when this is true.

When type I error is committed, the conclusion for the Test product is BIOEQUIVALENT, when in fact it is not. This may have as a consequence that patients consume a non-interchangeable formulation. During hypothesis testing, this risk is set at 5% ($\alpha = 0.05$).

The Type II error: This is encountered when the null hypothesis is accepted and the latter is false. That is, the conclusion is NON-BIOEQUIVALENCE, when it really is. Consequently, a formulation is penalized that would otherwise be on the market. This risk is dependent on the study design and its performance, and is identified as (β).

Therefore, statistical power is defined as the probability of rejecting the null hypothesis when this is false. This means that there is the probability of considering two formulations as NON-BIOEQUIVALENT, when they are really NO-BIOEQUIVALENT.

Power is defined as:

$$\text{Power (P)} = 1 - \beta \qquad \text{(Eq. 12)}$$

Obviously, there is dependence between power and the sample size necessary to reach the predefined power [9].

Let us once again consider the example of Clopidogrel and the following equation:

$$n_e = \left[t\left(\alpha/2, 2n-2\right) + t\left(\beta, 2n-2\right)\right]^2 \left[\frac{\sigma_d}{\Delta}\right]^2 \qquad \text{(Eq. 13)}$$

If a significance level of 0.05 is desired during hypothesis testing, then $\alpha/2 = 0.05/2 = 0.025$, and the degrees of freedom (2n-2) deriving from two sequences with 14 volunteers per sequence are 26; thus, it is necessary to seek the value of $t_{(\alpha/2,\, 2n-2)}$ in the t-distribution.

$$t(0.975, 26) = 2.056$$

Similarly, assuming β as 0.2, we have:

$$t(1-\beta, 2n-2) = t(0.8, 26) = 0.856$$

Finally, the term σ_d is the ratio of the coefficient of variation of previously reported data [8], and is divided by the maximum difference expected between bioequivalent formulations (20%).

$$\sigma_d / \Delta = 59.38/20 = 2.969$$

Thus, the number of volunteers required in each sequence (n_e) is:

$$n_e = (2.056+0.856)^2 (2.969)^2 = 74.84$$

This means 75 subjects per sequence for a total 150 study volunteers.

INTER- AND INTRA-SUBJECT VARIABILITIES

As can be observed in Table 8, the lack of statistical power in the Clopidogrel study was due to the wide range of responses obtained in each subject on each product (treatment).

Therefore, estimation of the ABE must additionally consider the variability in the bioavailability; the ABE takes into consideration a priori that the variances of both treatments are equal:

$$\sigma_T^2 = \sigma_R^2$$

where:

σ_T^2 is the Test product variability, and

σ_R^2 is the Reference variability

If Test product variability is greater than that of the Reference product, the opinion of interchangeability can be questioned. An example of this situation can be noted in Figure 4.

Test product variability can be calculated and must fall within Reference-product variability. This estimation can be effected based on an interval of a certain percentage. Such a percentage exceeds the therapeutic window of the drug.

Prior to initiating variability calculations, it is necessary to keep the following items in mind [7]:

σ_s^2 is the Inter-subject Variability (Between), and is calculated as follow:

$$\sigma_s^2 = \frac{MS\,inter - MS\,intra}{2} \quad \text{(Eq. 14)}$$

σ_e^2 is the Intra-subject variability (Within), calculated as follows:

$$\sigma_e^2 = MS\,intra = \frac{SS\,intra}{n_1 + n_2 - 2} \quad \text{(Eq. 15)}$$

In order to evaluate the equality of variances between Test and Reference formulations, it is necessary to consider that each volunteer has two different responses, one for Test and another for Reference product. Such responses are defined as Y_{iRk} and Y_{iTk}.

Variability for these responses can be represented as:

$$Var(Y_{iTk}) - Var(Y_{iRk}) = (\sigma_T^2 + \sigma_s^2) - (\sigma_R^2 + \sigma_s^2) \quad \text{(Eq.16)}$$

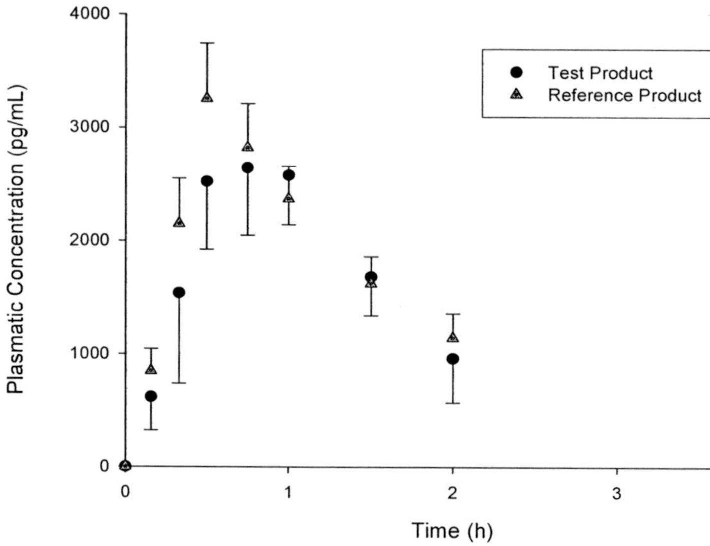

Figure 5. Variability of the Test and Reference products

The variation of each formulation is within the Inter-subject variability. On simplifying Eq. 16, it is possible to observe that:

$$Var(Y_{iTk}) - Var(Y_{iRk}) = \sigma_T^2 - \sigma_R^2 \qquad \text{(Eq. 17)}$$

One of the most frequently employed methods for evaluation of product variance is the Pitman-Morgan procedure, which assumes the following hypothesis:

$$H_0 : \sigma_T^2 = \sigma_R^2 \text{ vs. } H_a : \sigma_T^2 \neq \sigma_R^2$$

supposing that carryover and period effects are absent, and visualizing Eq.17 SS:

$$Var(Y_{iTk}) = S^2{}_{TT} \qquad \text{(Eq. 18)}$$

and

$$Var(Y_{iRk}) = S^2{}_{RR} \qquad \text{(Eq. 19)}$$

Then, we have:

$$S_{TT}^2 = \frac{1}{n_1 + n_2 - 1} \sum_{k=1}^{2} \sum_{i=1}^{n_k} (Y_{iTk} - \bar{Y}_T)^2 \qquad \text{(Eq. 20)}$$

$$S_{RR}^2 = \frac{1}{n_1 + n_2 - 1} \sum_{k=1}^{2} \sum_{i=1}^{n_k} (Y_{iRk} - \bar{Y}_R)^2 \qquad \text{(Eq. 21)}$$

$$S_{TR} = \frac{1}{n_1 + n_2 - 1} \sum_{k=1}^{n} \sum_{i=1}^{n_k} (Y_{iTk} - \bar{Y}_T)(Y_{iRk} - \bar{Y}_R) \qquad \text{(Eq. 22)}$$

The test to evaluate similarities between variances (F_{PM}) is calculated as follows:

$$F_{PM} = \frac{(n_1 + n_2 - 2)[F_{TR} - 1]^2}{4F_{TR}(1 - r_{TR}^2)} \qquad \text{(Eq. 23)}$$

where:

$$r_{TR} = \frac{S_{TR}}{\sqrt{(S_{TT}^2)(S_{RR}^2)}} \qquad \text{(Eq. 24)}$$

$$F_{TR} = \frac{S_{TT}^2}{S_{RR}^2} \qquad \text{(Eq. 25)}$$

On replacing the Pitman-Morgan statistic:

$$F_{PM} = \frac{(n_1 + n_2 - 2)(F_{TR} - 1)^2}{4F_{TR}(1 - r_{TR}^2)} \qquad \text{(Eq. 26)}$$

H_0 is rejected, and it is concluded that variances of Test and Reference products are different only when the following condition is valid:

$$F_{PM} > F(\alpha, 1, n_1 + n_2 - 2)$$

Using Table 3 data, it is possible to observe:

S^2_{tt}	S^2_{rr}	S_{tr}
39517078.9	26469269.4	16037454.3

F_{tr}	1.49294181
r_{tr}	0.4958748
F_{PM}	2.96881015
P	0.09082684

The conclusion, that is, entertaining a probability >0.05, comprises acceptance of the null hypothesis, establishing equality of variances between both products.

With the data of the example of Clopidogrel, four conclusions can be established:

1. There is not sufficient statistical power to reach a conclusion of bioequivalence.
2. To reach minimum power, it is necessary to have a 150-volunteer sample size.
3. Reference product variability is greater than that corresponding to Test product variability.
4. Normal data distribution is assumed solely for logarithmic data.

It is obvious that a study in which 150 volunteers are required for to obtain valid conclusions is quite cumbersome in practice. Control of a high number of subjects is complex, and the cost is increased tremendously.

Because of all the factors mentioned previously. Novel experimental designs that allow for ease of increasing observations without a high number of subjects have been adopted. Such designs, evaluate in a dissected manner, Within- and Between- subject variations.

According to FDA 2001 guidelines (8), the approach for HVD and HVDP is through Individual Bioequivalence (IBE), which considers higher-order designs (in which the number of compared formulations is less than the number of sequences or periods).

ANALYSIS OF VARIANCE FOR REPEATED MEASUREMENTS

When there is more than one factor influencing the evaluated response, it is logical to consider that there may be interactions among all these different factors.

Interactions are dependent on the response produced by the level of certain factor on the levels of other factors. This phenomenon cannot be understood as an additive process.

Each studied factor can be: a) A fixed effects factor, or b) a Random effects factor.

A fixed effects factor has pre-defined levels. For example, in a crossover design, Period is defined from the beginning as levels 1, 2,… j. Sequence is another fixed effects factor, which has two previously assigned levels (TR and RT); in the case of higher-order designs, these can be TRT and RTR.

Regarding higher-order designs, these are very useful because they permit unbiased calculation of Intra-subject variance. In these designs, each subject takes at least one product twice (2x3), or both products twice (2x4). These designs are shown in Tables 9–13.

Table 9. FDA Proposal, 2x3 Crossover design

SEQUENCE	PERIOD I	PERIOD II	***PERIOD III***
RTR	*R*	*T*	*R*
TRT	*T*	*R*	*T*

Table 10. 2x3 Two-sequence dual design

SEQUENCE	PERIOD I	PERIOD II	***PERIOD III***
RTT	*R*	*T*	*T*
TRR	*T*	*R*	*R*

Table 11. 2x3 Extra-Reference Design

SEQUENCE	PERIOD I	PERIOD II	***PERIOD III***
RTR	*R*	*T*	*R*
TRR	*T*	*R*	*R*

Table 12. 2x4 Crossover Design

SEQUENCE	PERIOD I	PERIOD II	PERIOD III	***PERIOD IV***
RTRT	*R*	*T*	*R*	*T*
TRTR	*T*	*R*	*T*	*R*

Table 13. 4X2 Balaam Design

SEQUENCE	PERIOD I	**PERIOD II**
TT	*T*	*T*
RR	*R*	*R*
TR	*R*	*T*
RT	*T*	*R*

Higher-order designs use the following mathematical model [11, 13]:

$$Y_{ijk} = \mu + Q_k + S_{ik} + P_j + F_{jk} + C_{(j-1k)} + W_{jkl} + e_{ijk} \qquad \text{(Eq. 27)}$$

where:

Y_{ijk} = is the response (ABC_{0-t}, ABC_{0-inf} or Cmax) of ith subject in k^{th} sequence in j^{th} period.

μ = the overall mean.

Q_k = the fixed effect of kth sequence $(\Sigma_k Q_k = 0)$

S_{ikl} = the randomized effect of ith subject in k^{th} sequence with l^{th} formulation.

P_j = the effect of jth period $(\Sigma_j P_j = 0)$

$F_{(j,k)}$ = the effect of formulation in k^{th} sequence that has been administered in j^{th} period. $(F_R+F_T = 0)$

$C_{(j-1k)}$ = the first-order carryover effect of formulation in k^{th} sequence that is administered in (j-1) period $(C_R+C_T = 0)$

e_{ijk} = the residual error.

W_{ljk} = the interactions of fixed effects $(\Sigma_l W_{ljk} = \Sigma_j W_{ljk} = \Sigma_k W_{ljk} = 0)$

(S_{ikR}, S_{ikT}) are independent randomized vectors with identical distribution, with zero as mean and unknown co-variance matrix:

$$\begin{pmatrix} \sigma_{BT}^{2} & \rho\sigma_{BT}^{2}\sigma_{BR}^{2} \\ \rho\sigma_{BT}^{2}\sigma_{BR}^{2} & \sigma_{BR}^{2} \end{pmatrix}$$

e_{ijk}s are random errors with zero as mean and variance σ^2_{Wl} (F=R or F=T) (S_{ikR}, S_{ikT}) are not dependent on e_{ijk}s.

Note that σ^2_{BT} and σ^2_{BR} are Between-subject variances, while σ^2_{WR} and σ^2_{WT} are Within-subject variances.

ρ (rho) is the correlation between the responses of the same measurement in the same subject for both Test and Reference formulations.

The main differences between higher-order models and standard 2x2 designs comprise the inclusion of the fixed effect of sequence, the randomized effects of subject, and the possible interactions. The fixed effects of sequence allow for evaluation of the randomized drug-dosing sequence.

Data obtained from higher-order models are examples of repeated measurements. Information that these models offer with respect to measurements of Between- and Within-subject variations are of great interest in the analysis of this type of assay.

Between-subject information is contained in the total measurements of the entire population (mean), while Within-subject information is contained in the total of measurements in the same volunteer. The latter gives rise to use of models that consider individual subject-effects in addition to population-effects (mixed models) [14].

Introduction of S_{jkl} as a random effect modifies the co-variance structure of observed responses, having:

$$V(Y_{ijk}) = V(\varepsilon_{ijk}) + V(S_{ijk}) = \sigma_{Wl}^{2} + \rho\sigma_{Bl}^{2} \qquad \text{(Eq. 28)}$$

where:

$l = R, T$ and

$$Cov(Y_{ijk}, Y_{ij'k}) = \rho\sigma_{BR}\sigma_{BT} \quad j \neq j' \qquad \text{(Eq. 29)}$$

Due to that measurements of treatments applied to one subject are dependent, the usual methodology for estimation of least squares and the theory that supports it, fail. In such a case, an algorithm known as Restricted Maximum Likelihood (REML) is applied.

REML-generated equations cannot be resolved analytically for each component of variance. It is necessary to obtain solutions by means of iterative methods.

Running mixed models for repeated measurements can be carried out employing PROC MIXED (SAS), *lme* (S-PLUS) or R, among other several software packages.

Mixed models possess two main advantages when analyzing crossover designs:

- More precision on estimation of treatment-effects by using the information obtained from repeated measurements of the same subject.
- Drop-out of subjects during this type of study is quite common. If subjects were treated as fixed-effects, then measurement of subjects with missing data could be used. However, if subjects are treated as random-effects, then all measurements, even those derived from incomplete observations, can be used during analysis. It is assumed that repeated measurements from each subject obtain from a probabilistic distribution.

INDIVIDUAL BIOEQUIVALENCE (IBE)

The key in selecting IBE lies in comparing the difference of variances between Test and Reference products, with the corresponding difference of variances between Reference product and itself.

The ratio of these differences is denominated "Individual Difference Ratio" (IDR). To determine whether two products are individually bioequivalent, both Reference and Test formulations must be administered more than once to the same subject, thus, it is logical to conduct replicated design studies. If IDR are ca, 1 (0.8–1.25), both products can be considered Individually Bioequivalent.

Regarding the Eq. 27 model:

$$Y_{ijk} = \mu + Q_k + S_{ik} + P_j + F_{jk} + C_{(j-1k)} + W_{jkl} + e_{ijk}$$

$$Var(S_{ijT} - S_{ijR}) = \sigma_{BT}^2 + \sigma_{BR}^2 - 2\rho\sigma_{BR}\sigma_{BT}$$

$$= (\sigma_{Bt} - \sigma_{BR})^2 + 2(1-\rho)\sigma_{BR}\sigma_{BT} \quad \text{(Eq. 30)}$$

This variance obtains from the interaction subject-formulation and is denoted by σ_D^2

According to FDA 2001 guidelines [4], the way to judge individual bioequivalence is through the following criterion:

$$\theta = \frac{\delta^2 + \sigma_D^2 + \sigma_{WT}^2 - \sigma_{WR}^2}{max(0.04, \sigma_{WR}^2)} \quad \text{(Eq. 31)}$$

where:

σ_D^2 is the interaction subject-formulation variance.

δ is the difference of the means (μ_T-μ_R) between formulations.

The way to judge individual bioequivalence is through the following test:

$$H_0: \theta \geq \theta_U \quad vs. \; H_1: \theta < \theta_U$$

where:

θ_U is the upper limit for bioequivalence established by FDA 2001 guidelines.

Sample Size for Higher-Order Designs

Considering the Schuirmann test (a two one-sided hypothesis assay), the power function is defined as:

$$P_{k(\theta)} = F_{vk}`[\frac{(\Delta-\theta)}{C.V.\sqrt{b_k / n}}] - t(\alpha, v_k) - F_{vk}(t\alpha, v_k) - \left[\Delta + \theta / C.V.\sqrt{b_k / n}\right] \quad \text{(Eq. 32)}$$

where:

$$\theta = 0$$

Then, sample size calculation will be:

$$n \geq b_k\left[t(\alpha, \nu_k) + t(\beta/2, \nu_k)\right]^2\left[\frac{C.V.}{\Delta}\right]^2 \quad \text{(Eq. 33)}$$

if:

$$\theta > 0$$

Then:

$$n \geq b_k\left[t(\alpha, \nu_k) + t(\beta/2, \nu_k)\right]^2\left[\frac{C.V.}{\Delta - \theta}\right]^2 \quad \text{(Eq. 34)}$$

where:

$CV = \sqrt{\exp(\sigma^2) - 1}$

σ^2 is the Within- residual variance in log-scale.
K = 1,2,3,4 (number of sequences)

and

$\nu_1 = 4n\text{-}3$

$\nu_2 = 4n - 4$

$\nu_3 = 6n - 5$

$\nu_4 = 12n - 5$

For the term b:

$b_1 = 2$
$b_2 = 3/4$
$b_3 = 11/20$
$b_4 = 1/4$

Eq. 34 has been used for calculating sample size with the objective of reaching a power of 80% and detecting a 20% difference between formulations for the three higher-order designs previously discussed [9].

Table 14. Number of subjects required in Schuirmann test to reach a power of 80%, considering $\Delta = 0.2$, $\alpha = 0.05$, AND $\theta = 100x\ (\mu T - \mu R)/\mu R$ ***(Chow SC 2009)***

DESIGN	C.V. (%)	θ	
		5%	10%
Balaam	24	132	288
	30	200	448
	40	356	796
Two-sequence dual	24	18	38
	30	38	86
	40	68	150
2x4 design	24	14	28
	30	28	62
	40	50	110

In Table 14 the number of volunteers required to reach 80% power is compared by Schuirmann test, considering a CV between 24 and 40% and θ of 5 and 10%.

We note that for the Balaam design, the number of subjects is higher compared with that of two-sequence dual and 2x4 designs. On the other hand, when dual design is compared with 2x4, we observe that the number of subjects is not significantly different.

Also noteworthy is that these designs are additionally useful for ABE.

Discussion and Conclusions

As could be fully evidenced for the classic 2x2 design, it is impossible to evaluate Between-subject variability, which in HVD and HVDP is very high, in isolation. In addition, in this design type, the carryover effect is contained within the sequence effect; thus, in cases with a sequence effect, dissection of the analysis is rendered very difficult.

With the advent of higher-order designs, it was feasible to evaluate each of the sources of variation in bioequivalence trials. One type of these designs, the Balaam model, allows evaluating the carryover design in isolated fashion. This model considers that the cited effect is not the same in sequence TR as that in mirror sequence RT.

According to FDA guidelines for establishing bioequivalence in HVD and HVDP, it is strongly required to employ higher-order designs for the isolated evaluation of Within- and Between-subject variation, the latter with the aim of demonstrating Individual Bioequivalence based on the construction of acceptance criteria supported by the width of the Reference variance σ^2_{WR}. Therefore, any design that allows the estimation of σ^2_{WR} will possess great advantages over the classic 2x2 design.

In FDA guidelines, two experimental designs to evaluate IBE are described: 2x3, and 2x4.

The 2x4 design presented in these guidelines has been widely discussed by several authors [3,9,10,13]. This design has been considered as very complete, because Between-subject variability, Within-formulation variability (σ^2_{WR}, σ^2_{WT}), and carryover effect are evaluated simultaneously.

However, in practice, the 2x4 design is complicated. It has some disadvantages, such as:

1. The clinical period is too lengthy, considering that each volunteer needs to participate in four periods. This may promote drop-outs.
2. Consecutive doses in the same volunteer may increase the risk of incurring side effects.

The 2x3 design proposed by the FDA guidelines has some advantages over 2x2, such as:

(a) It can evaluate Within-formulation variability and Between-subject variability simultaneously.
(b) TR and RT carryover effects can be evaluated.

However, compared with the 2x4 design, the 2x3 design possesses two main disadvantages:

- It cannot be assumed that RT and TR carryover effects are the same. Thus, this design cannot evaluate the carryover effect wisely.
- The number of subjects required to reach power is higher.

Therefore, other authors have proposed two 2x3 designs: The Extra-Reference period, and the Dual two-sequence design (depicted in Tables 10 and 11). These two designs have major advantages over the FDA guidelines' 2x3 proposal, such as:

1. They evaluate the carryover effect completely.
2. They evaluate the direct-formulation effect.
3. The Reference variance is estimated.
4. They utilize a repeated measurement design.

Additionally, these two designs permit having same number of observations from the Reference product as from the 2x4 design and estimating σ_{WT}^2 with the same precision.

According to Table 14, 2x3 Dual design requires the same number of volunteers as the 2x4 design to reach the same statistical power.

Moreover, the Extra-Reference 2x3 design represents a better IBE assay, considering that bioequivalence limits are fixed bases on Reference product variance.

Finally, the 3x3 crossover design (RTR, TRR, and RRT), which does not present major advantages over 2x3 Dual or Extra Reference designs, has been proposed recently.

Future Perspectives

The more frequent use of higher-order designs will drive the simplest experimental designs in terms of degree of complexity, length of clinical periods, and costs.

We consider that continuous use of 2x3 Dual and Extra-Reference designs will demonstrate their robustness and practicality. This will bring about acceptance of these designs by regulatory authorities.

References

[1] Blume, HH; Elze, M; Potthast, H; Schug, S. Practical strategies and design advantages in highly variable drug studies: multiple dose and replicate administration design. In: HH; Blume, KK. Midha, (Eds.), Bio-International 2: Bioavailability, *Bioequivalence, and Pharmacokinetic Studies*, Medpharm, Stuttgart, Germany, 1995, 117-122.

[2] Steinijan,s, VW. Bioequivalence assessment –European perspectives. In: International Bioequivalence Standards: A New Era. G; Amidon, L; Lesko, K; Midha, V ; Shah, J. Hilfinger, (Eds.), TSRL Press, *Ann Arbor,*

MI, USA, 2006, 191-199.

[3] Midha, KK; Rawson, MJ; Hubbard, JW. The bioequivalence of highly variable drugs and drug products. *International Journal of Clinical Pharmacology and Therapeutics*, 2005, *43(10)*, 485-498.

[4] FDA (CDER). Guidance for Industry. Statistical Approaches to Establishing Bioequivalence. FDA (CDER). *US Department of Health and Human Services*, Rockville, MD, USA. January, 2001.

[5] Jones, B; Kenward, MG. Higher-order designs for two treatments. In: Design and Analysis of Cross-Over Trials. Chapman & Hall (Eds.), *CRC Press, Boca Raton*, FL, USA, 2003, 138-169.

[6] Secretaría de Salud. NOM-177-SSA1-1998. Pruebas y procedimientos para demostrar intercambiabilidad de formulaciones farmacéuticas. *Secretaría de Salud*, México, 1998, 53.

[7] Chow, SC; Liu, JP. Statistical Inferences for Effects from a Standard 2x2 Crossover Design. In: Design and Analysis of Bioavailability and Bioequivalence Studies. (3rd ed.), B; Jones, JP; Liu, KE; Peace, Bw. Turnbull, (Eds.), Chapman & Hall/CRC Biostatistics Series, *Boca Raton*, FL, USA, 2009, 55-75.

[8] Kumar, RT; Usha, PR; Naidu, MUR; Gogtay, JA; Meena, M. Bioequivalence and Tolerability Study of Two Brands of Clopidogrel Tablets, Using Inhibition of Platelet Aggregation and Pharmacodynamic Measures. *Current Therapeutic Research*, 2003, *64(9)*, 685- 696.

[9] Chow, SC; Liu, JP. Optimal crossover designs for two formulations for average bioequivalence. In: Design and Analysis of Bioavailability and Bioequivalence Studies. (3rd ed.), B; Jones, JP; Liu, KE; Peace, BW. Turnbull, (Eds.), Chapman & Hall /CRC Biostatistics Series, *Boca Raton*, FL, USA, 2009, 257-293.

[10] Haidar, SH; Davit, B; Cheng, ML; Conner, D; et al.: Bioequivalence approaches for highly variable drugs and drug products. *Pharmaceutical Research*, 2008, *25(1)*, 237-241.

[11] Cheng, ML; Lesko, LJ. Individual bioequivalence revisited. *Clinical Pharmacokinetics*, 2001, *40(10)*, 701-706.

[12] Tothfalusi, L; Endrenyi, L; Midha, KK; Rawson, MJ; Hubbard, JW. Evaluation of the bioequivalence of highly-variable drugs and drug products. *Pharmaceutical Research*, 2001, *18(6)*, 728 -733.

[13] Chow, SC; Shao, J; Wang, H. Individual bioequivalence testing under 2X3 designs. *Statistics Medicine*, 2002, 21, 629-648.

[14] Brown, H; Prescott, R. Cross-Over Trials. In: Applied Mixed Models in Medicine. John Wiley & Sons, *Ltd.*, 2006, 271-308.

[15] Senn, S. Some special designs. In: *Cross-Over Trials in Clinical Research*, 2nd ed. John Wiley & Sons, 2002, 205-243.

In: Biomedical Chromatography
Editors: John T. Elwood, pp. 105-140
ISBN: 978-1-60741-291-5

Chapter 4

NEW STRATEGIES OF SAMPLE PREPARATION IN THE ANALYSIS OF PHARMACEUTICALS AND THEIR DETERMINATION USING HIGH PERFORMANCE LIQUID CHROMATOGRAPHY

Zoraida Sosa-Ferrera, Cristina Mahugo-Santana, MA Esther Torres-Padrón and José Juan Santana-Rodríguez*
University of Las Palmas de Gran Canaria,
Las Palmas de Gran Canaria, Spain

ABSTRACT

Pharmaceuticals are "emerging contaminants" in the environment, which have received increasing attention in recent years, in part, because these compounds are constantly being emitted into the environment in quantities similar to other organic contaminants, and thus may be candidates for future regulation. This paper reviews recently published data with respect to sample-preparation techniques for determining

* Department of Chemistry. Faculty of Marine Sciences. University of Las Palmas de Gran Canaria. 35017. Las Palmas de Gran Canaria, Spain. E-mail: jsantana@dqui.ulpgc.es

pharmaceutical traces in various different environmental matrices using liquid chromatography. An overview of multi-residue analytical methods, covering sample extraction and purification as well as chromatographic separation and various different detection methods is provided to illustrate common trends and method variability. Additionally, recent developments and advances in this field are presented.

1. Introduction

Over the last few years, increasing attention has been given in analytical chemistry research to the development of methodology that allows for detection of all the substances in the environment which may represent a threat to our health. The global manufacture of chemical products has increased enormously over the last few decades, in general in benefit of humankind although certainly not lacking in toxic side-effects, since they are highly persistent and, thus, can accumulate dangerously in organisms (Daughton et al., 1999; Halling-Sorensen et al., 1998).

Nowadays, many of our problems of pollution are due to intermittent or continuous leakage of these substances into our environment. Besides their aforementioned properties of toxicity, persistence and dangerous bioaccumulation, there is already clear evidence of the impact that these substances may have on the biological processes of flora and fauna, and even on human beings.

Of all the substances within this category, pharmaceutical compounds have become of increasing concern in the last two decades. The use of pharmaceutical compounds has accelerated mainly as the result of the increased life expectancy of the population in the developed world. Many pharmaceutical compounds have been developed without their effects on the environment being fully understood or researched. Thousands of tons of pharmaceuticals are used every year in human and veterinary medicine throughout the world and enter the environment through metabolic excretion and/or improper disposal. Some compounds are not completely eliminated in sewage treatment plants, and enter the environment through a variety of sources and pathways (Ternes, 1998).

These compounds are designed to specific purposes and many are programmed to persist in the body. These features, among others, have made pharmaceuticals into the objective of evaluation for their potential effects on aquatic organisms (Sanderson et al., 2004) and non-target species (Cleuvers, 2003; Laville et al., 2004; Fent et al., 2006).

Pharmaceutical concentrations measured in surface waters are generally well below concentrations known to cause acute toxicity in aquatic organisms. However, chronic exposure to pharmaceuticals has the potential for numerous other subtle effects, such as metabolic or reproductive changes in non-target organisms (Daughton and Ternes, 1999). Some recent studies have shown that pharmaceuticals with highly specific mechanisms, can elicit profound effects at extremely low concentrations. It is clear that aquatic life is sensitive to at least some pharmaceutical compounds. Much more research is needed to establish whether the aquatic environment is susceptible to exposure to pharmaceuticals. Although individual concentrations of any one drug may be low, it must be considered that combined concentrations of drugs sharing common mechanisms may be substantial. Therefore, monitoring is required to provide greater knowledge with respect to their occurrence in the environment, to understand how they behave, how they distribute and what are the effects when organisms are exposed to low levels (Halling-Sorensen et al., 1998).

Although little is known of the occurrence and effects of pharmaceuticals in the environment, more data exist for antibiotics than for any other therapeutic type of drug or medicine. This is a result of their extensive use in both human and veterinary therapy.

The influx of antibiotics into the environment is of particular interest because bacteria exposed to antibiotic traces may induce resistance (Daughton and Ternes, 1999; Chee-Sandford et al., 2001). There is indirect evidence that exposure to antibiotics has produced resistance of bacteria in the environment (Witte 1998; Guardabassi et al., 1998). Antibiotic residues in the environment are also a risk to microbial organisms responsible for beneficial processes, in estuarine and marine environments, such as denitrification, nitrogen fixation and organic breakdown (Costanzo et al., 2004).

Both the number of pharmaceutical compounds licensed for human use and their annual consumption have increased dramatically. In Spain, the compounds licensed for human use have increased approximately by 21,8% in the last two decades (Memoria de la Agencia española de medicamentos y productos sanitarios 2006). These compounds and their metabolites have the potential to filter into the environment. Despite this fact, there is no monitoring of the level of contamination caused by such pollutants in most environmental samples.

The quantification of pharmaceuticals in human biological matrices such as blood, plasma or urine has been developed for a long time (Snow, 2000).

Nevertheless, the quantification of pharmaceuticals in complex environmental samples presents greater difficulties, due to the fact that these compounds are present in very low concentrations and there are multiple different types of compounds.

GC and LC are the most widely and frequently used techniques to monitor the concentrations of contaminants in environment (Hernando et al., 2006; Petrovic et al., 2002). When target compounds and the degradation products are too polar, non-volatile or thermodegradable compounds, they cannot be analysed by GC, thus making LC into an essential tool for the determination of this type of compounds (Kolpin et al., 2002; Kümmerer, 2004).

One feature that distinguishes pharmaceutical compounds from other recognized persistent organic pollutants is the higher polarity of the parent pharmaceuticals. This, and the consideration that these substances are present in the environment at trace concentrations and in complex matrices, makes it necessary for suitable preparation techniques to be applied to the samples in order to isolate and preconcentrate the analytes prior to their determination. Thanks to extraction protocols that have been developed and coupled with more sensitive analytical methods, the concentrations of some drugs have been detected in environmental medium such as water, solid and sludge.

An ideal sample preparation protocol should be simple, cost effective, fast and reproducible. Sample preparation must ensure rapid, quantitative and selective extraction of the active pharmaceutical compounds. Another significant feature is the possibility of parallel multi-sample pretreatment, in order to reduce total analysis time.

Although many traditional sample preparation methods are still in use, such as, liquid-liquid extraction (LLE) and Soxhlet extraction, they present several drawbacks, such as complicated, time-consuming procedures, large amounts of sample and organic solvent and difficult automation. That is why the trends in recent years have been toward replacing it by other methods that use small volumes of organic solvent or none, thus increasing the selectivity of the extraction.

The most common technique, in the case of determination of pharmaceuticals in environmental samples, is solid-phase extraction (SPE) (Stumpf et al., 1999; Hilton and Thomas, 2003; Roberts and Bersuder, 2006; Miao et al., 2002; Brown et al., 2006) together with solid-phase microextraction (SPME) (Carpinteiro et al., 2004; Lord and Pawliszyn, 2000) for water matrices, and ultrasonic solvent extraction (USE), pressurized liquid extraction (PLE) plus microwave assisted extraction (MAE) for solid samples.

This overview covers the methodologies developed in recent years for sample-preparation techniques used in the analysis of pharmaceutical compounds with differing characteristics in environmental samples using liquid chromatography. Additionally, there is a description of recent developments in the use of micellar media in this field.

2. Sample Pre-Concentration Procedures for Water Samples

The concentration levels of pharmaceuticals found in environmental water samples are generally too low to allow direct determination by liquid chromatography (LC). Therefore, despite its high sensitivity, a pre-concentration step is normally necessary to admit the low limits of detection (LODs) required in the analysis of environmental waters.

Sample preparations can be prepared using a wide range of techniques, but all the methods share the same goal (Smith, 2003):

- to remove potential interferences;
- to increase the concentration of an analyte;
- if necessary, to convert an analyte into a more suitable form; and,
- to provide a robust and reproducible method that is independent of variations in the sample matrix.

Few analytical approaches are based on the classical liquid–liquid extraction (LLE) (Soliman et al., 2004). The main drawback of this method is the requirement of large amounts of high-purity solvents that are expensive and toxic, resulting in the production of hazardous waste, the disposal of which is problematic (Psillakis and Kalogenakis, 2002. The technique is also time-consuming, tedious, and laborious.

This method has been gradually replaced by SPE and SPME, since these methods offer a series of advantages over the previous strategies:

1. higher recoveries;
2. improved selectivity, specificity and reproducibility;
3. elimination of emulsions;
4. less organic solvent usage;
5. shorter sample preparation time; and,

6. easier operation plus the possibility of automation.

2.1. Solid Phase Extraction

Solid phase extraction (SPE) is widely used as a preparation process based on its selective retention of the compounds of interest in a solid sorbent that can then be eluted with an organic solvent. Figure 1 shows a scheme of the SPE procedure.

The choice of sorbent is the key point in SPE because it controls parameters such as selectivity, affinity and capacity. This choice depends strongly on the analytes of interest and the interaction of the selected sorbent through the functional groups of the analytes, as well as on the kind of sample matrix and its interaction with both the sorbent and the analytes (Fontanals et al., 2005).

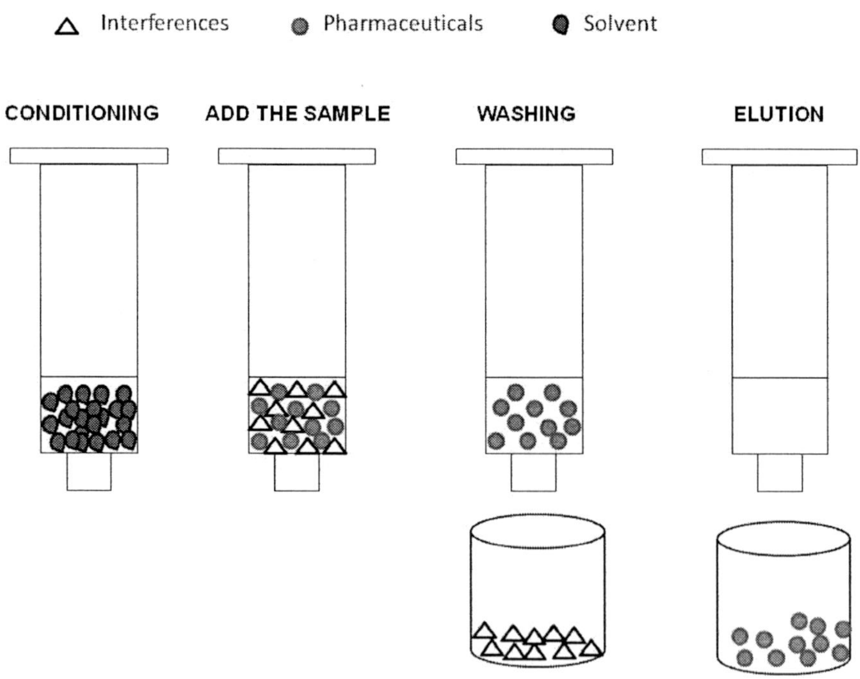

Figure 1. Scheme of SPE procedure.

Several SPE materials have been used for the extraction, pre-concentration and clean-up of pharmaceutical residues in water, from the conventional alkyl-modified silica materials (C-18 non-polar phase) to the new materials based on polymer sorbents that improve the retention for polar compounds. The OASIS HLB has high rates of recoveries. This cartridge exhibits both hydrophilic and lipophilic retention characteristics, and has often been used for simultaneous extraction of neutral and acid pharmaceutical traces (Kolpin et al., 2002). Multi-residue methods for different types of pharmaceuticals using OASIS-HLB at neutral pH have also been reported recently by Barceló and co-workers (Gómez et al., 2006; Gros et al., 2006; Petrovic et al., 2006).

Although the OASIS HLB seems to be the favourite material for multi-class pharmaceutical residue analysis, other materials have also been used. For example, in a recent study for the determination of nine basic pharmaceuticals in wastewater and surface water, Speedisk phenyl cartridges that allow for identification and quantification of these pharmaceuticals in the ng/l and µg/l-range were used (Van De Steene and Lambert, 2008).

For specific groups of pharmaceuticals different cartridges have been applied. Strata-X was selected as the best phase for extracting sulphonamides, tetracyclines, fluoroquinolones, penicillin G procaine and trimethoprim in mixes using off-line SPE (Roberts and Bersuder, (2006; Hilton and Thomas, 2003; Mutavdzic et al., 2006).

In some cases, two cartridges have been used in tandem. For example, an anion-exchange cartridge followed by an OASIS HLB cartridge have been used in the detection of fluoroquinolones, sulphonamides and trimethoprim in wastewater samples. The extracted antibiotics were analyzed using liquid chromatography with electrospray mass spectrometry (Renew and Huang, 2004).

However, many interfering species may co-elute on conventional sorbents. Due to this limitation, sorbents with higher selectivity such as immunosorbents and molecularly imprinted polymers (MIPs) have been developed and applied to different samples. MIPs are synthetic polymers with highly specific recognition of target molecules (Dickert, 2007). They have been used in fields where a certain degree of selectivity is required, and can solve all problems of extraction, especially in the case of polar compounds. Nowadays the use of MIPs in SPE, so-called molecularly imprinted SPE (MISPE), is one of the most advanced technical applications of these polymers (Baggiani et al., 2007; Tamayo et al., 2007). This approach has been applied to the extraction of various different pharmaceutical compounds by several research groups (Caro et al., 2005; Hu et al., 2005), and offers a very promising type of sorbent, although time-consuming, and requiring patience and skill in handling.

The SPE of pharmaceuticals is often done off-line. However, the technique is well-suited to on-line procedures and automation, which may be considered one of the most advantageous sample preparation techniques available. The SPE cartridge can be installed in the injection valve instead of the injection loop and the preconcentrated analytes directly eluted onto the analytical column. The main advantages of this technique are minimum sample manipulation, automation, less time-consuming and improved throughput (Kuster et al., 2006; Pozo et al., 2006; Rodríguez-Mozaz et al., 2007). Other important advantages of on-line coupling are the decreased risk of contamination of the sample or sample extract, elimination of analyte losses by evaporation or by degradation during sample pre-concentration together with improved precision and accuracy. The analysis of whole samples leads to lower limits of detection and smaller sample volumes needing to be used to obtain adequate sensitivity for a wide variety of compounds (Rodríguez-Mozaz et al., 2007). Thus, there are numerous examples of the application of on-line SPE followed by LC in the analysis of pharmaceutical compounds (Pitarch et al., 2004; Bones et al., 2006; Quintana et al., 2006).

2.2. Solid Phase Microextraction

The demand for a reduction in the quantity of organic solvents used in SPE has led to the use of other solvent-free techniques, such as solid-phase microextraction (SPME). Solid phase microextraction was developed by J. Pawliszyn (Arthur and Pawlisyzn, 1990) and has attracted the interest of researchers worldwide over the last decade. SPME was originally developed for use with gas chromatography (GC), but was subsequently successfully coupled to liquid chromatography (SPME–LC) for the analysis of semi- and non-volatile compounds such as drugs (Chen and Pawliszyn, 1995).

The method is based on a partition equilibrium of the analyte between the aqueous phase and a short piece of a fused silica fiber coated with a polymeric stationary phase. The extraction of the analytes from the aqueous sample can be carried out via direct immersion of the fiber into the sample (DI-SPME) or by headspace exposure (HS). The advantage of the latter approach is that the matrix in the sample cannot interfere with the fiber, but can only be used for volatile compounds.

Although SPME was initially considered only for the analysis of organic compounds in environmental and water samples, it is now being used more and more in other fields to determine proteins, polar alkaloids, pharmaceuticals and surfactants, because of its successful coupling with liquid chromatography (LC), capillary electrophoresis (CE) and mass spectrometry (MS) devices (Abdel-Rehim et al., 2000; Theodoridis et al., 2000; Vas and Vekey, 2004). The sensitivity and precision are generally good or better than standard methods, and the methods themselves are simpler, while solvent use is eliminated or minimized. The SPME method has proved to be a viable technique for overcoming the matrix effects of environmental samples.

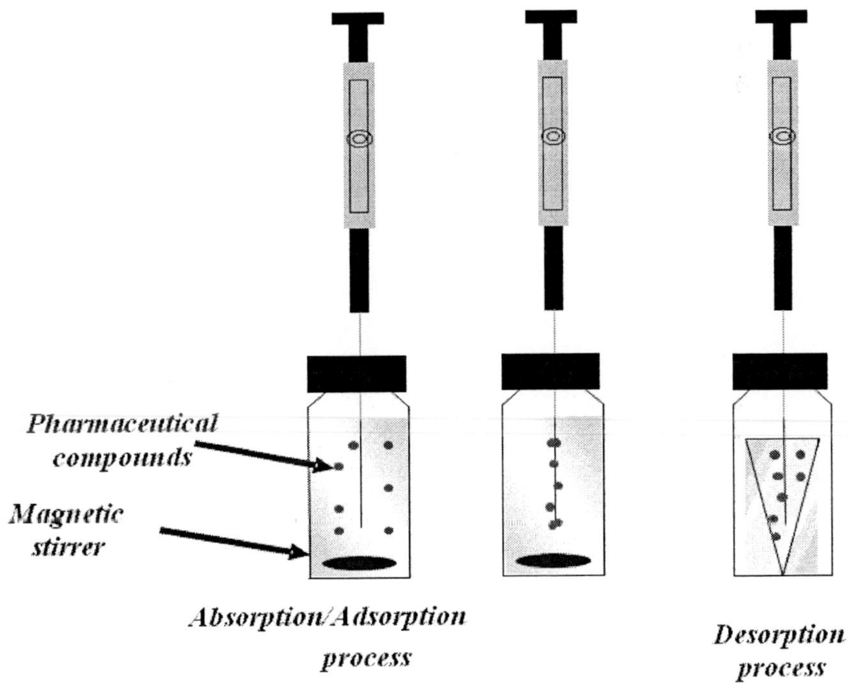

Figure 2. Scheme of SPME procedure.

Figure 2 illustrates a layout of a typical SPME procedure. The heart of the SPME analysis is the fiber. Historically, in the late 1990s, "first-generation" fibers were introduced, such as polydimethylsiloxane (PDMS) and polyacrylate (PA). The former offers great affinity in the extraction of non-polar compounds, whereas the latter, PA is polar and more suitable for extracting polar compounds. However, both phases have linear structures (the polymer layer is not cross-linked), and therefore are quite fragile and unstable, leading to fiber breakage, film stripping and severe injector contamination. Nowadays, a wide array of SPME fiber coatings are commercially available, increasing the range of compounds that may be extracted. These coatings have been blended with divinylbenzene (DVB) and carbowax (CW), such as PDMS/divinylbenzene (PDMS/DVB), Carbowax/DVB (CW/DVB) and Carboxen/polydimethylsiloxane (CX/PDMS), which present larger specific surface areas and greater potential for extracting polar compounds. These fibers have proved to be more stable, allowing for a larger number of applications before their performance begins to fall off.

The practical application of the SPME method requires for optimum performance of the main experimental parameters involved in the process, such as the type and thickness of the fiber, the extraction time, the extraction temperature and other features such as agitation, pH and ionic strength of the sample, together with the characteristics which affect the desorption process, such as time and composition of the desorption solvent. The repercussions of these effects are discussed in length by H. Lord and J. Pawliszyn (2000).

SPME has offered broad-ranging applications for different types of organic pollutants in several different kinds of matrices. Nevertheless, its use for pharmaceutical extractions has been limited to biological matrices such as milk, urine and plasma (Theodoridis et al., 2004; Oliveira et al., 2005; Salgado-Petinal et al., 2005; Musteata and Pawliszyn, 2007). Its application in environmental matrices is much less frequent, although SPME has lately been increasingly used as sample-preparation technique for isolating pharmaceuticals and their metabolites in environmental samples (Moeder et al., 2000; Lamas et al., 2004; Rodríguez et al., 2004; Zorita et al., 2007; Vera-Candioti et al., 2008).

Balakrishnan et al. (2006) compared DI-SPME with SPE procedure for extracting sulphonamides from wastewater. SPE was not effective in determination of sulphasalazine (not detectable after SPE) as opposed to SPME, which allows for the extraction of all sulphonamide compounds with an efficiency above 75% (except sulphamethazine (39.8%) and sulphamethoxazole (59.2%)). The same paper described thc optimization of the SPME method and its application in the determination of target pharmaceuticals.

A similar study was carried out by McClure and Wong (2007). They presented an SPME method for the simultaneous extraction, from wastewater, of several macrolide, sulphonamide, and trimethoprim antibiotics frequently to be detected in environmental waters, while also comparing the SPME-derived measurements with those obtained using SPE, to assess the advantages and limitations with regard to the SPME method.

Solid-phase microextraction (SPME) has also been tested in the extraction of tetracyclines (TCs), from water (Lock et al., 1999). In this study, the optimisation of the on-line SPME–LC–MS method is described including choice of extracting fiber and desorption method (heating or salting out the analytes).

In order to achieve maximum efficiency in the extraction of organic compounds and improvements in the chromatograms, a new desorption mode has been presented, which uses a micellar medium as a desorbing agent, in SPME combined with HPLC (Mahugo Santana et al., 2007; Torres Padrón et al., 2008). Surfactants are capable of solubilising various different kinds of solutes. In particular, non-ionic surfactants have been widely used in the extraction of various different organic substances from varying types of matrices, thereby demonstrating their great potential as extractants. Moreover, they are compatible with the aqueous-organic mobile phase in chromatographic analysis, which facilitates these applications. The use of surfactants allows us to establish environmentally friendly methods without the use of organic solvents, which are sources of contamination.

This new approach, the so-called solid phase microextraction with micellar desorption, (SPME-MD) has been applied to the determination of pharmaceutical residues in water samples (Montesdeoca Esponda et al., 2009; Torres Padrón et al., 2009). In these studies, the research was aimed at optimising the two phases of the SPME process, including the type of fiber to be used together with the surfactant in the desorption step. In both studies, it has been demonstrated that micellar desorption improves upon the results obtained with conventional SPME and methanol as desorbing agent.

In the specific case of detection of fluoroquinoles (FQ) in liquid samples (Montesdeoca Esponda et al., 2009), the best conditions for the analysis were: direct exposure of the CW/TPR fiber, and micellar desorption with Polyoxyethylene 10 lauryl ether (POLE) for 15 min. Figure 3 shows the chromatograms obtained for a fluoroquinolone mixture with conventional desorption using methanol (a) and SPME-MD with POLE (b). We can observe that micellar desorption improves the response of all the fluoroquinolones significantly over the results with methanol desorption.

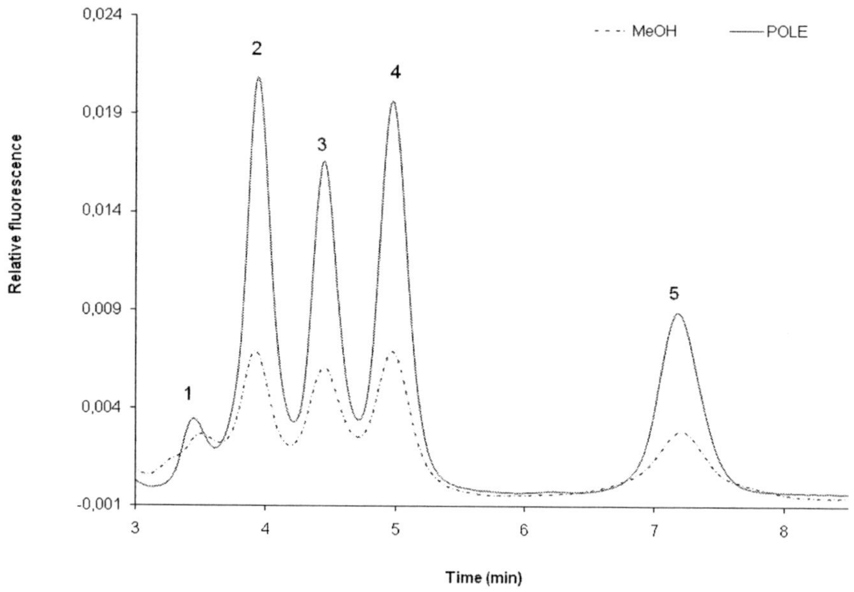

Figure 3. Chromatograms obtained for a fluoroquinolones mixture with 50 ng mL^{-1} for (1) Levofloxacin (2) Norfloxacin (3) Ciprofloxacin (4) Enrofloxacin (5) Sarafloxacin with conventional desorption using methanol (a) and SPME-MD with POLE (b) (Montesdeoca Esponda et al., 2009).

Torres Padrón et al. (2009) describe the development, validation and application of an SPME-MD process followed by HPLC-DAD for the simultaneous analysis of six non-steroidal acidic anti-inflammatory drugs (NSAIDs) and lipid regulators in water samples of differing origins. A 65 μm PDMS-DVB fiber and two surfactants, POLE and $C_{12}E_6$ were selected to obtain high rates of efficiency in the SPME-MD process for these target analytes. Micellar desorption was compared with the conventional desorption method using off-line organic solvent desorption. In general, analyte responses are lower and apolar analytes have worse resolution with a broadening of the chromatographic peaks when conventional desorption with methanol was used. Micellar desorption is observed to improve the response of the studied analytes (Figure 4).

High-throughput applications and automated instrumentation are becoming more and more important. In-tube SPME (Kataoka, 2002; Bagheri and Salemi, 2004) has been developed mainly to advance in this direction, with stir bar sorptive extraction (SBSE) (Baltussen et al., 1999) designed to increase the sensitivity of the analysis.

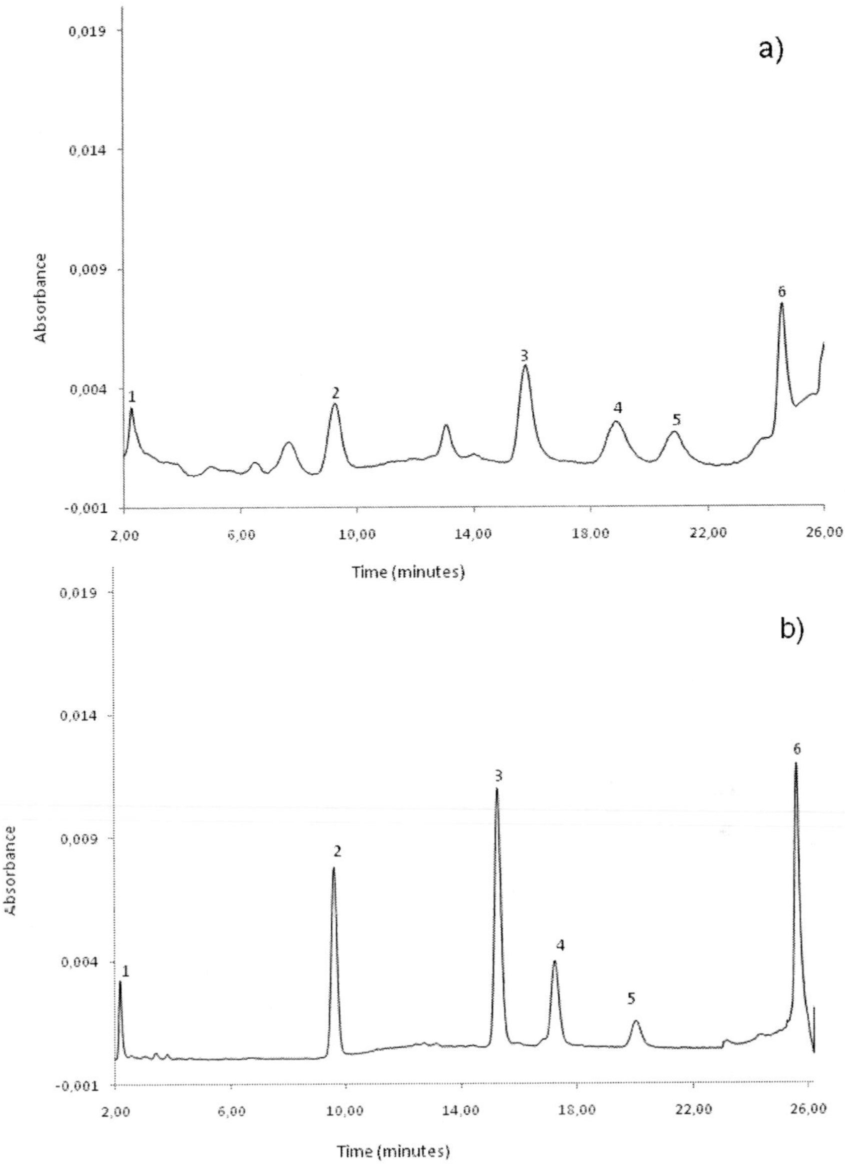

Figure 4. Chromatograms of a groundwater sample spiked with 500 ng mL^{-1} for (1) Phenazone (2) Carbamazepine (3) Ketoprofen (4) Naproxen (5) Bezafibrate and (6) Ibuprofen. a) Chromatogram obtained by direct injection; b) Chromatogram obtained by SPME-MD using 2.5% (v/v) $C_{12}E_6$ as desorbing agent (Torres Padrón et al., 2009)

In-tube SPME requires more complex instrumentation than regular SPME, but higher sensitivity can be obtained by using a longer tube and, consequently, more sorbent. In this approach, the analytes are directly extracted from the sample into the coated stationary phase of a capillary column and then they are desorbed by introducing a stream of mobile phase or static desorption solvent when the analytes are more strongly adsorbed to the capillary coating. This technique is also suitable for automation which is a great advantage in the coupling of SPME to HPLC systems. Automated SPME-HPLC will facilitate shorter total analysis time, better accuracy and higher precision than manual techniques.

Wen et al. (2006) have developed an on-line method for simultaneous determination of four endocrine disruptors (17β-estradiol, estriol, bisphenol A and 17α-ethinylestradiol) in environmental waters by coupling in-tube solid-phase microextraction (SPME) to high-performance liquid chromatography (HPLC) with fluorescence detection (FLD). A poly(acrylamide-vinylpyridine-*N*,*N'*-methylene bisacrylamide) monolith was selected as the extraction medium. Low detection and quantification limits were achieved in the range of 0.006–0.10 ng/mL and 0.02–0.35 ng/mL for spiked lake waters, respectively.

Determination of five fluoroquinolones (FQs), (enoxacin, ofloxacin, ciprofloxacin, norfloxacin, and lomefloxacin) in environmental waters, using a fully automated method with in-tube solid-phase microextraction (SPME) coupled with liquid chromatography–tandem mass spectrometry (LC/MS/MS) was developed by Mitani and Kataoka (2006). Carboxen 1010 PLOT capillary column was used by way of an extraction device. The detection limits obtained for the five FQs ranged from 7 to 29 pg/mL. The in-tube SPME method showed 60–94-fold higher sensitivity than the direct injection method (5 μL injection). This method was applied successfully to the analysis of environmental water samples without any other pretreatment and interference peaks.

Baltussen and co-workers (1999) designed a novel approach based on an extraction mechanism similar to SPME, to improve the extraction efficiencies and the amount of extracted analytes. In this technique, called stir-bar sorptive extraction or SBSE, a stir bar, coated with PDMS, is used. The advantage of SBSE is that there is a much larger mass of PDMS available, resulting in higher enrichment factors and capacity. The volumes of PDMS typically found with an SPME and an SBSE are 0.5 μl and 25 μl, respectively. Sampling of large volumes can be performed in a relatively short time using a stir bar. The SBSE has been widely used for enrichment and sensitive determination of

Table 1. Determination of pharmaceutical compounds in water samples.

Analytes	Samples	Extraction	Determination	Analitycal Parameters	Ref.
NSAIDs	River and tap water	On-line SPE Chromolith C18 monolithic silica cartridge Acetonitrile and water at pH 4	LC-UV-ESI-MS	Recoveries: 9-107% RSD < 20% LODs: 0.008-0.165 $\mu g{\cdot}L^{-1}$	J. Bones et al.
NSAIDs (Ibuprofen, naproxen, ketoprofen, tolfenamic acid, diclofenac)	Sewage water	SPME PA fiber, pH 3, 30 min, room temperature	GC–MS	RSDs: 4-9% LOQs: 12-40 $ng{\cdot}L^{-1}$	I.Rodríguez et al.
Sulfonamides, macrolides, and trimethoprim	Wastewater	SPME CW/TPR fiber, room temperature, 30 min, NaCl 30% (w/v), at pH 4 and 7	HPLC–ESI-MS/MS	RSDs: 7-50% LODs: 0.08-6.1 $ng{\cdot}L^{-1}$	E.L. McClure et al.
Anti-epileptic, analgesic, anti-inflammatory, antidepressants, anti-ulcer and ß-blockers drugs	Hospital effluent wastewater	SPE Oasis HLB cartridges at pH 7 Elution with methanol	LC–ESI-MS/MS	Recoveries: 45-114% RSDs:0.3-4.9% LODs:7-47 $ng{\cdot}L^{-1}$	M.J. Gomez et al.
Multi-class pharmaceuticals	Surface and wastewater	SPE Oasis HLB cartridges Elution with methanol	LC–ESI-MS/MS	Recoveries: 50 -116% RSDs<11% MDLs: 1-160 $ng{\cdot}L^{-1}$	M.Gros et al.
SSRIs	River water and sewage water	SPME PDMS-DVB fiber at 100 ◦C, 30 min NaCl 15%, acetylation before extraction	GC–MS	Recoveries:88-120% RSDs:1.9-14.7% LODs: 0.017-0.075 $ng{\cdot}L^{-1}$	J.P.Lamas et al.
Basic pharmaceuticals	Wastewater and surface water	SPE Speedisk phenyl cartridges at pH 7 Elution with methanol NH_2 column for clean-up	LC–ESI-MS/MS	Recoveries: 60–100% RSDs: 3-15% LODs: 0.05–1 $ng{\cdot}L^{-1}$	J.C.Van De Steene et al.
Fluoroquinolones	Surface water and wastewater	In-tube SPME Carboxen 1010 PLOT column at pH 8	LC-ESI-MS/MS	Recoveries:82-97% RSDs:1.9–8.6% LODs:7-29 $pg{\cdot}mL^{-1}$	K.Mitani et al.
Biologically active polar substances	Wastewater	SPME PA fiber, 30 min, pH 2, NaCl	GC–MS	RSDs:10-18% LODs:0.2 and 50 $\mu g{\cdot}L^{-1}$	M.Moeder et al.
Analgesics, antiinflammatories, psychiatric drugs, anti lcer agents, antibiotics, ß-blockers	Wastewater and river water	SPE Oasis HLB cartridges Elution with methanol	UPLC-Q–TOF–MS	RSDs: 0.5-9.1% MDLs: 10-500 $ng{\cdot}L^{-1}$	M.Petrovic et al.

Table 1. (Continued)

Analytes	Samples	Extraction	Determination	Analitycal Parameters	Ref.
Sulphametoxazole, bezafibrate, metoprolol, carbamazepine and bisoprolol	Ground, surface and drinking-water	-	cLC–MS–MS	Recoveries: 50 - 114% RSDs below 20% LODs: 10-20 ng·L^{-1}	E.Pitarch et al.
Antibiotics	Ground and surface water	SPE C_{18} cartridge. Elution with ammonium acetate at pH 2.5	LC–MS/MS	Recoveries: 74% - 123% RSDs < 14% LODs: 0.4–4.3 ng·L^{-1}	O.J.Pozo et al.
Fluoroquinolones, sulfonamides and trimethoprim	Wastewater samples	SPE HLB cartridges. Elution with MeOH and H_3PO_4	LC–MS	Recoveries: 37- 129% MDLs: 2- 90 ng·L^{-1}	J.E.Renew et al.
OSPAR priority pharmaceuticals	Surface water and treated sewage effluent samples	SPE Sstrata-X, 200 cartridges. Elution with methanol, methanol/acetic acid and methanol/ ammonium hydroxide	HPLC-ESI-MS/MS	Recoveries: 10 - 95% RSDs: 4 - 13% LODs : 1-20 ng·L^{-1}	P.H.Roberts et al.
Fluoroquinolones	Seawater, sewage,and groundwater	SPME-MD CW-TPR fiber, 60 min. at 40 °C,NaCl (30% w/v). Desorption with POLE,	HPLC-FLD	Recoveries: 81 - 116% RSDs < 9% LODs: 0.01–0.2 ng·mL^{-1}	S.Montesde oca et al.
Acidic pharmaceuticals	River, sea and wastewater	SBSE PU and PDMS fibers	HPLC-DAD	Recoveries: 10 – 91% RSDs < 15% LODs: 0.40–1.7μg·L^{-1}	A.R.M.Silva et al.
Estrogens	Tap water, lake water and sewage water	in-tube SPME Poly(acrylamide-vinylpyridine-*N*,*N*_-methylene bisacrylamide) monolith, synthesized inside a PEEK tube at pH 6	HPLC–FLD	Recoveries: 86 - 116% RSDs: < 12% LODs: 0.006–0.10 ng·mL^{-1}	Y.Wen et al.

Abrev. : NSAAID: non-steroidal acidic anti-inflammatory drugs; SSRIs: Selective serotonin reuptake inhibitors; OSPAR: Oslo and Paris Commission for the protection of the Marine Environment of the North East Atlantic; LC: liquid chromatography, GC: Gas chromatography, HPLC: high performance liquid chromatography; UPLC: Ultra pressure liquid chromatography;UV: Ultraviolet visible detection, FLD: fluorescence detection; DAD: Diode array detection; MS: mass spectrometry; ESI: Electrospray ionization; cLC: capillary liquid chromatography; SPME-MD: Solid-phase microextraction with micellar desorption; SBSE: Stir bar sorptive extraction; RSD: Relative standard deviation; MDL: Method detection limit; LOC: limit of confirmation, LOD: limit of detection.

priority organic micro-pollutants in water samples, as well as in other matrices (Nakamura and Daishima, 2005; Nakamura and Daishima, 2004; Juan-Garcia et al., 2005; Duran Guerrero et al., 2006; Blasco et al., 2004).. Almeida and Nogueira (2006) have used SBSE and liquid chromatography with diode array detection (SBSE–LD–HPLC/DAD) to simultaneously detect several steroid sex hormones (SSHs) in water and urine matrices.

However, this analytical approach was inspired by the use of polydimethylsiloxane (PDMS), which is a polymeric phase with higher affinity for non-polar compounds, and with certain limitations when attempting to extract more polar compounds. Recently, a novel polymeric phase based on polyurethane (PU) foams has been proposed (Neng et al., 2007). It has been demonstrated that these polymers present remarkable stability and excellent mechanical resistance, and are highly recommendable in the case of enrichment of more polar analytes in aqueous media (Portugal et al., 2008).

The capacity and potential application of PU foams as compared to the conventional PDMS during SBSE was evaluated for the determination of six non-steroidal acidic anti-inflammatory drugs (NSAIDs) and lipid regulators in water matrices. The performance of the proposed methodology was evaluated in terms of precision, linearity and detection limits. The best detection limits achieved with PU foams are a consequence of the better recovery yields obtained with this phase, due to higher affinity with more polar compounds (Silva et al., 2008).

Table 1 summarizes SPE and SPME procedures for the extraction of pharmaceuticals from water samples.

3. Sample Preconcentration Procedures for Solid Samples

Residues of pharmaceutical compounds end up in the environment. Conventional wastewater treatment processes are not specifically designed to remove pharmaceuticals, so they often do not eliminate them efficiently. It is only recently that attention has been given to the potential contamination caused by these pharmaceuticals residues in soil and sediment samples. The large surface area and activity of soils, sediments and suspended soil matter contribute towards the accumulation of these compounds in abiotic solid samples, allowing them to persist in the environment, possibly affecting aquatic and earth organisms that depend on these substrates.Pharmaceuticals are large and chemically complex molecules. The wide range of chemical categories represented in this group means that generalisations with respect to their behaviour is impossible, and that their mobility in soil and sediment can also vary (Oppel et al., 2004; Beausse, 2004).

The traditional sample-preparation method for solid phases, the Soxhlet extraction method, has largely been replaced by other methods such as ultrasonic solvent (USE) (López de Alda et al., 2002; Löffler and Ternes, 2003; Ternes et al., 2005), pressurized liquid extraction (PLE) (Dıáz-Cruz et al., 2006; Yang and Metcalfe, 2006; Jacobsen and Halling-Sørensen, 2006; Stoob et al., 2006) and microwave assisted extraction (MAE) (Morales et al., 2005), used to enhance extraction efficiency. Moreover, these techniques present other characteristics such as short extraction time, decreased solvent consumption and decreased sample handling.

Sample extracts obtained from solid matrices may occur with interfering coextracts, which demand additional clean-up before their analysis. In this case, SPE is the most commonly used method chosen for sample clean-up in that it allows large sample volumes to be concentrated and purified in one step.

A number of factors must be considered when a method is being chosen to extract pharmaceutical compounds and their offshoot degradation products from soil samples. For samples with a matrix as complex as soil, it is vital to consider the dominant sorptive mechanisms when designing a successful and efficient extraction technique.

Table 2 gives references for the USE, PLE and MAE procedures for the extraction of pharmaceuticals from solid environmental samples.

3.1. Ultrasonic Extraction (USE)

Ultrasonic Extraction (USE) is often used for the extraction of pharmaceuticals from solid samples. The sample, with added solvent, is immersed in an ultrasonic bath and subjected to ultrasonic radiation for a few minutes. Extracted analytes are separated from the matrix by vacuum filtration or centrifuging. The process is repeated twice or three times to achieve higher extraction efficiency, and the extracts are combined for analysis. USE has the benefit of shorter extraction times, thus allowing for several extractions to be performed simultaneously and requiring no specialised laboratory equipment.

Table 2. Determination of pharmaceutical compounds in solid samples.

Analytes	Samples	Extraction	Clean-up/ preconcentration	Determination	Analitycal Parameters	Ref.
NSAIDs Ibuprofen Naproxen Ketoprofen Diclofenac	River sediment samples	MAE Methanol/acetone	SPE StrataX reversed-phase cartridges	GC-MS	Recoveries: 46 -87% RSDs: 6 - 11% LODs: 0.03 - 0.08 $\mu g \cdot g^{-1}$	J.Antonic´ et al.
Phenazone, carbamazepine, clofibric acid, ketoprofen, naproxen, bezafibrate, ibuprofen and propranolol	Sediment samples	MAME POLE	SPE Oasis HBL cartridges	HPLC–UV–DAD	Recoveries over 70% RSDs < 11% LODs: 4 -167 $ng \cdot g^{-1}$	R. Cueva – Mestanza et al.
Sulphonamides and penicillins	Sludge samples	PLE Methanol/acetone	SPE Oasis HBL cartridges	LC–ESI–MS/MS	Recoveries : 14–104% RSDs < 20% LODs: 1 pg/g-0.2 $ng \cdot g^{-1}$	M.S. Díaz-Cruz et al.
Sulfonamides and macrolide antimicrobials	Sewage sludge samples	PLE Water/methanol	SPE Oasis HLB cartridges	LC-MS/MS	Recoveries: 78-142% RSDs: 2-20% LOQs: 3- 41 $\mu g \cdot kg^{-1}$	A.Göbe et al.
Tetracyclines, sulfonamides, and tylosin	Swine manure	PLE Methanol/ citric acid	SPE Tandem SAX-HLB cartridges	LC–ESI–MS/MS	Recoveries > 70% LOQs: 10–100 μg kg−1	A.M.Jacobsen et al.
Estrogens and progestogens	River sediments	USE Methanol/acetone	SPE Sep-Pak Plus C-18 cartridges	LC-DAD-MS	Recoveries: 64-102% RSDs: 6-19% LODs:0.04–1.00 $ng \cdot g^{-1}$	M.J.López de Alda et al.
Triclosan, 2,4-DCP and 2,4,6-TCP	Sludge and sediment samples	MAE Methanol/acetone	SPE Oasis HLB cartridges	GC–MS/MS	Recoveries: 78-106% RSDs < 13% LOQs: 0.4-0.8 $ng \cdot g^{-1}$	S. Morales et al.
Macrolides, sulfonamides, ranitidine, omeprazole and trimethoprim	Sewage sludge samples	PLE Water (pH 3):methanol	SPE Oasis HLB cartridges	HPLC–MS	Recoveries: 54 - 95% RSDs < 15% LODs: 2 - 11$\mu g \cdot kg^{-1}$	A.Nieto et al.

Table 2. (Continued)

Analytes	Samples	Extraction	Clean-up/ preconcentration	Determination	Analitycal Parameters	Ref.
Antibiotics, acidic pharmaceuticals and prescribed Medications	Wastewater treatment plants samples	Ultrasonic Extraction USE Methanol/acetone	SPE Oasis HLB cartridges	LC-ESI-MS/MS	Recoveries: 22 - 105% LOQs: 0.17-8.11 $ng \cdot L^{-1}$	A.L.Spongberg et al.
Sulphonamides	Aged soil samples	PLE Water/Acetonitrile	-	LC–MS/MS	Recoveries: 41 - 93% RSDs < 10% LODs < 15 $\mu g \cdot kg^{-1}$	K.Stoob et al.
Alkaloids	*Rhizoma Coptidis*	MAME Genapol X-080	CPE	HPLC–UV	Recoveries: 94 – 95 % RSDs < 3.3%	C.Sun et al.
Acidic pharmaceuticals, antibiotics and ivermectin	River sediment samples	USE Methanol, acetic acid, acetone and ethyl acetate	SPE Oasis MCX cartridges and LiChrolute EN glass cartridges	LC–MS/MS	LOQs: 0.4 - 20 $ng \cdot g^{-1}$	D.Löffler et al.

Abrev. : NSAAID: non-steroidal acidic anti-inflammatory drugs; LC: liquid chromatography, GC: Gas chromatography, HPLC: high performance liquid chromatography; UPLC: Ultra pressure liquid chromatography;UV: Ultraviolet visible detection, FLD: fluorescence detection; DAD: Diode array detection; MS: mass spectrometry; ESI: Electrospray ionization; cLC: capillary liquid chromatography; MAE: microwave assisted extraction; MAME: microwave assisted micellar extraction POLE: Polyoxyethylene 10 lauryl ether; CPE: cloud point extraction; PLE: pressurized liquid extraction; USE: Ultrasonic Extraction; SPE: solid phase extraction; RSD: Relative standard deviation; MDL: Method detection limit; LOC: limit of confirmation; LOQ: limit of quantization; LOD: limit of detection.

The technique is relatively inexpensive compared to most modern extraction methods. The main disadvantages of USE are its poor reproducibility values due to lack of uniformity in the distribution of ultrasound energy, together with the fact that it is not easily automated and is not suitable for volatile analytes.

A number of papers have been published dealing with ultrasonic- assisted extraction of various different pharmaceutical residues. Turiel et al. (2006) have developed a method for the simultaneous analysis of several quinolones and fluoroquinolones in soil samples based on extraction by USE and quantification by HPLC/UV. The optimum extraction procedure is based on the formation of antibiotic-Mg(II) complexes, which allow for desorption and extraction of both groups of antibiotics in a single step, something which is not feasible when using a conventional organic solvent. The detection limits achieved are suitable for the determination of quinolones and fluoroquinolones in soil samples at realistic environmental concentration levels. Spongberg and Witter (2008) have presented a method for the analysis of twenty compounds including several classes of antibiotics, acidic pharmaceuticals and prescribed medications, using a combination of ultrasonic extraction, SPE cleanup and liquid chromatography–electrospray ionization tandem mass spectrometry in three wastewater treatment plant.

3.2. Pressurized Liquid Extraction (PLE)

Pressurized Liquid Extraction (PLE), also called accelerated solvent extraction (ASE), extracts solid samples under high pressure and at high temperatures. These aspects increase the extraction efficiency and rate, while reducing the consumption of organic solvents and operating time. The high temperatures used in PLE diminish the surface tension and the viscosity of the extractant. Thus, the solvent penetrates better and further into the interstitial spaces of the soil sample, thereby increasing the solubility of the analytes.

However, the use of elevated temperatures may entail serious disadvantages since there may be collateral coextraction of unwanted soil matrix components and thermal degradation of the target analytes. The co-extracted matrix poses significant problems in analyte detection and requires intensive clean-up steps in order to eliminate or reduce interferences prior to analysis. SPE is preferred in most instances because it is fast, requires a low volume of organic solvent and has a low risk of contamination.

PLE has significant advantages over other methods, such as better reproducibility, minor use of extraction solvent and minor total time of sample preparation. The extracts are generally much more concentrated than with other conventional extraction methods.

PLE is a technique with multiple applications in environmental chemistry and food analysis. In all the cases here presented, the application used was a combination of pressurized liquid extraction, solid phase and liquid chromatography with tandem mass spectrometry (PLE-SPE-LC-MS/MS). For example, Göbel et al. (2005) optimized a method for the determination of sulphonamides, macrolides, and trimethoprim in sewage sludge; Barron et al. (2008) published an analytical method for quantifying 27 pharmaceutical residues in soil and treated sludge, at $ng.g^{-1}$ levels. Likewise, Jelic et al. (2009) used the same approach for sample preparation in the simultaneous analysis of 43 pharmaceutical residues in sewage sludge and sediment samples.

In some cases, there was no SPE clean-up step, and filtered PLE extracts were directly analyzed, as in the method of Nieto et al. (2007) for the detection of 11 pharmaceuticals in sewage sludge samples using PLE and LC-MS with positive electrospray ionization.

Okuda et al. (2009) studied the occurrence of 66 PPCPs (pharmaceuticals and personal care products) in sewage sludge samples. To decide the optimal extraction conditions, two extraction methods, ultrasonic solvent extraction (USE) and pressurized liquid extraction (PLE) methods, were used. An appropriate procedure including PLE using water (pH 2), PLE using methanol (pH 4), and USE using a mixture of methanol and water (1/9,v/v, pH 11) was found to be most effective in that the total recovery of most of the PPCPs indicated 40 to 130%

3.3. Microwave Assisted Extraction (MAE)

Microwave Assisted Extraction (MAE) involves heating the solid sample and solvent in a closed vessel with microwave energy under controlled temperature and pressure conditions. This closed extraction system accelerates the extraction process and improves the yield performance of the method. It has been shown to significantly reduce both the time and amount of solvent used. The main parameters influencing MAE performance include: solvent type, solvent volume, microwave power and exposure time and temperature. The extraction solvents used are polar liquid or mixtures of polar and non-polar liquids because only polar solvents are able to absorb microwave energy.

The solvent volume depends on the type and the size of sample, but small solvent volumes in the range of 10–30 ml are used on average. These volumes are about 10 times less than those used in classic extractions. The microwave power and the corresponding time depend on the type of sample and solvent used. Generally, extraction times in MAE are much shorter than those of classic extraction techniques with typical extraction procedure taking 15–30 min. Usually elevated extraction times do not improve extraction efficiency and, in some cases, may even decrease analyte recoveries due to the fact that thermolabile substances may decompose at high temperature.

MAE has been successfully applied to the extraction of environmental pollutants from solid samples and has also been used in environmental sample analysis to estimate the presence of pharmaceuticals and personal-care products (PPCPs). In most cases, an additional clean-up step is required prior to the analysis.

Dobor et al. (2009) have presented a sample preparation procedure for the determination of selected acidic pharmaceuticals (ibuprofen, naproxen, ketoprofen and diclofenac) in sewage sludge using microwave-assisted extraction followed by the conventionally applied solid phase extraction (SPE), before detection via gas chromatography–mass spectrometry. The recoveries calculated from the analytical data of spiked sludge samples were in the range of 80–105%±15% for the four pharmaceuticals.

Antonic and Heath (2007) have examined several extraction techniques (ultrasonic extraction, Soxhlet extraction, pressurized liquid extraction, supercritical fluid extraction and microwave assisted extraction) for the determination of the same pharmaceutical compounds in river sediment samples. They chose MAE as the most effective extraction method. This decision was based both on extraction efficiencies and standard deviations, together with average solvent consumption and extraction time per sample. While PLE and SFE use up to 80 mL of solvent and 50 mL of CO_2 per sample, MAE requires only 40 mL of solvent per sample. The extraction time for PLE and SFE is from 75 min to 90 min. per sample, whereas extraction time for MAE is 2 hours for 16 samples, averaging about 8 min. per sample.

An interesting alternative might be the introduction of surfactant solutions as extractants instead of organic solvents. This combination of MAE and micellar media as extractant gives the MAME technique. MAME has mainly been exploited in environmental analysis of solid matrices (Sosa Ferrera et al., 2004, Eiguren Fernández et al., 2001; Mahugo Santana et al., 2005), and recently also in isolating traditional Chinese drugs from plants (Sun and Liu, 2008, Sun et al., 2008).

This approach has been applied to the extract and preconcentrate of a selected group of eight pharmaceutical compounds belonging to various therapeutic categories from sediment samples (Cueva-Mestanza et al., 2008). A non-ionic surfactant, Polyoxyethylene 10 lauryl ether (POLE) was used for the MAME extraction, and SPE was used to clean-up and preconcentrate the target analytes in the extract, prior to their determination using HPLC–DAD. Relative recoveries for spiked sediment samples were over 70% and relative standard deviations (RSDs) were under 11% for all recoveries tested. Detection limits between 4 and 167 ng g^{-1} were obtained. Figure 5 shows a scheme of the MAME–SPE procedure with all of the various different specific steps represented. The optimised methodology was successfully applied to the analysis of target compounds in sediment samples of varying types and characteristics. In order to expand the use of micellar extraction techniques in this area of analysis, further exploration of other micelle-forming systems or mixed surfactant systems should be considered.

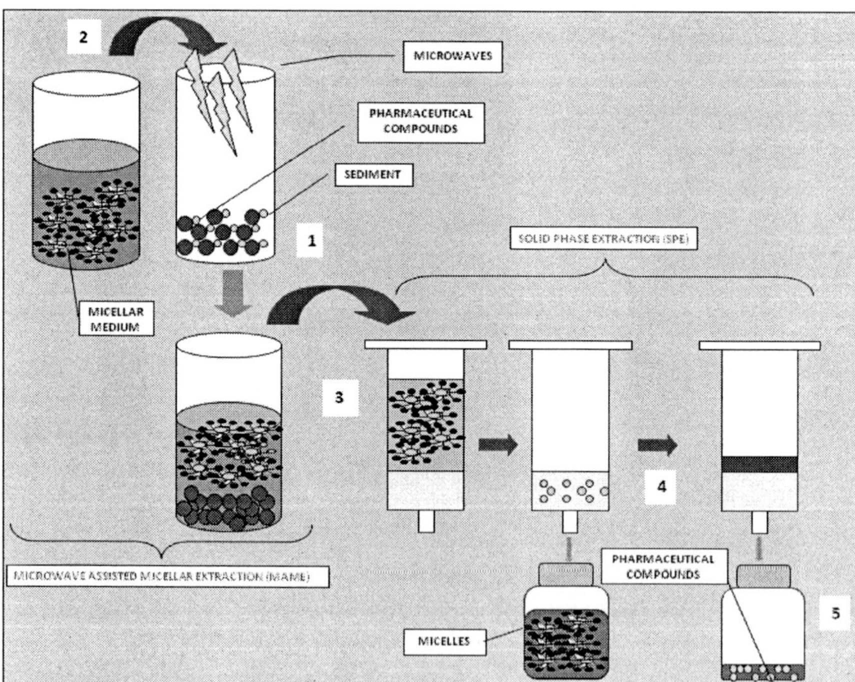

Figure 5. Scheme of a MAME–SPE procedure.

Conclusions and Trends

Over the last few decades, there has been growing awareness of the problem of the environmental pollution caused by pharmaceutical products. Two directives have been drawn up (the 2001/83EC for human pharmaceuticals, and the 2001/82/EC for animal pharmaceuticals) that demand environmental assessment for the approval of new drugs on the market. Preferentially, pharmaceuticals enter the environment through the wastewater treatment plants (WWTPs) effluents and land application of sludge as a result of the inadequacy of the methods of treatment for removal of the same. The assessment of contamination levels and new potential contaminants must address several aspects of this environmental problem. It is important to identify and determine the environmental concentrations of pharmaceuticals and their metabolites. In this respect, great efforts must be made to research new analytical methods for the assessment of pharmaceutical compounds in various different kinds of environmental matrices.

Several analytical approaches have been reported in the literature, but the majority of these have focused specifically on one family of pharmaceutical compounds. A significant challenge is met in the simultaneous extraction and analysis of multiple types of compounds, due to the wide range of polarities, solubilities, pKas and other properties under basic and acidic conditions. One of the main problems in monitoring the occurrence of pharmaceuticals in waste, or surface and ground waters is the lack of simple, sensitive and cost-effective analytical methods to quantify pharmacologically active substances (and their metabolites) in the concentration range of ng L^{-1}–μg L^{-1}. GC and LC are the techniques most used to monitor the concentrations of contaminants in the environment. When the target compounds and the degradation products are too polar, non-volatile or thermodegradable, they cannot be analysed using GC, thus making LC coupled to mass spectroscopy (LC–MS/MS) into the technique most frequently used, on account of its versatility, specificity and selectivity.

However, despite this high sensitivity, a preconcentratiom step is normally necessary to reach the low detection limits (LODs) required in the analysis of environmental samples. The fundamental step in this type of analysis is considered to be sample preparation. The research carried out to date has been directed at improving the extraction efficiency and selectivity through the development of new coated or parked materials, in the case of SPE and SPME, as new synthetic polymers suitable in specific separations. Meanwhile USE, PLE and MAE are adequate alternatives to the conventional Soxhlet extraction in the case of pharmaceutical compounds involving solid samples. The high extraction efficiencies reached with these techniques result in significant reductions in the solvent and time required for quantitative recovery of target compounds from complex matrices.

In order to expand the use of micellar extraction techniques in this area of analysis, further exploration of other micelle-forming systems or mixed surfactant systems should be considered. There are a large number of micelle-forming polymers in aqueous solution which have the potential to substitute the organic solvents used in classic extraction methodologies.

It is clear that advances in extraction, clean-up, separation and detection of pharmaceutical residues and their derivatives will be the key to obtain the information needed to assess the ecological risks and long term effects of these residues in the environment.

Acknowledgments

This work was supported by funds provided by the Spanish Ministry of Education of Science. Research Project N° CTQ 2006-06507.

References

Abdel-Rehim M, Bielestein M and Arvidsson T. Evaluation of solid-phase microextraction in combination with gas chromatography (SPME-GC) as a tool for quantitative bioanalysis. *Journal of Microcolumn Separation* 2000; 12: 308-315.

Almeida C and Nogueira JMF. Determination of steroid sex hormones in water and urine matrices by stir bar sorptive extraction and liquid chromatography with diode array detection. *Journal of Pharmaceutical and Biomedical Anaysis.* 2006; 41:1303-1311.

Antonić J and Heath E. Determination of NSAIDs in river sediment samples. *Analytical and Bioanalytical Chemistry* 2007; 387:1337–1342.

Arthur C and Pawlisyzn J. Solid phase microextraction with thermal desorption using fused silica optical fibers. *Analytical Chemistry* 1990; 62:2145-2148.

Baggiani C, Anfossi L and Giovannoli C. Solid phase extraction of food contaminants using molecular imprinted polymers. *Analytica Chimica Acta* 2007; 591: 29-39.

Bagheri H and Salemi A. Coupling of a modified in-tube solid phase microextraction technique with high performance liquid chromatography-fluorescence detection for the ultra-trace determination of polycyclic aromatic hydrocarbons in water samples. *Chromatographia* 2004; 59: 501–505.

Balakrishnan VK, Terry KA and Toito J. Determination of sulfonamide antibiotics in wastewater: A comparison of solid phase microextraction and solid phase extraction methods. *Journal of Chromatography A* 2006; 1131:1-10.

Baltussen E, Sandra P, David F and Cramers C. Stir bar sorptive extraction (SBSE), a novel extraction technique for aqueous samples: Theory and principles. *Journal of Microcolumn Separation* 1999; 11: 737–74.

Barron L, Tobin J and Paull B. Multi-residue determination of pharmaceuticals in sludge and sludge enriched soils using pressurized liquid extraction, solid phase extraction and liquid chromatography with tandem mass spectrometry. *Journal of Environmental Monitoring* 2008; 10:353–361.

Beausse J. Selected drugs in solid matrices: a review of environmental determination, occurrence and properties of principal substances. *Trends of Analytical Chemistry* 2004; 23:753-761.

Blasco C, Fernández M, Pico Y and Font G. Comparison of solid-phase microextraction and stir bar sorptive extraction for determining six organophosphorus insecticides in honey by liquid chromatography-mass spectrometry. *Journal of Chromatography A* 2004; 1030: 77–85.

Bones J, Thomas K., Nesterenko PN and Paull B. On-line preconcentration of pharmaceutical residues from large volume water samples using short reversed-phase monolithic cartridges coupled to LC-UV-ESI-MS. *Talanta* 2006; 70:1117-1128.

Brown KD, Kulis J, Thomson B, Chapman TH and Mawhinney DB. Occurrence of antibiotics in hospital, residential, and dairy effluent, municipal wastewater, and the Rio Grande in New Mexico. *Science of Total Environment* 2006; 366: 772-783.

Caro E, Marcé RM, Cormack PAG, Sherrington DC and Borrull F. Selective enrichment of anti-inflammatory drugs from river water samples by solid-phase extraction with a molecularly imprinted polymer. *Journal of Separation Science* 2005; 28:2080-2085.

Carpinteiro J, Quintana JB, Martınez E, Rodríguez I, Carro AM, Lorenzo RA and Cela R. Application of strategic sample composition to the screening of anti-inflammatory drugs in water samples using solid-phase microextraction. *Analytica Chimica Acta* 2004; 524:63-71.

Chee-Sandford J, Aminov R, Krapac I, Garrigues-Jeanjean N, Mackie R. Occurrence and diversity of tetracycline resistance genes in lagoon and groundwater underlying two swine production facilities. *Applied of Environmental Microbiology* 2001; 67:1494–502.

Chen J and Pawliszyn JB. Solid phase microextraction coupled to high-performance liquid chromatography. *Analytical Chemistry* 1995; 67:2530-2533.

Cleuvers M.. Aquatic ecotoxicity of pharmaceuticals including the assessment of combination effects. *Toxicological Letters* 2003; 142:185-194.

Costanzo SD, Murby J and Bates J. Ecosystem response to antibiotics entering the aquatic environment. *Marine Pollution Bulletin* 2004; 51:218–230.

Cueva-Mestanza R, Sosa-Ferrera Z, Torres-Padrón ME and Santana-Rodríguez JJ. Preconcentration of pharmaceuticals residues in sediment samples using microwave assisted micellar extraction coupled with solid phase extraction and their determination by HPLC–UV. *Journal of Chromatography B* 2008; 863: 150-157.

Daughton CG and Ternes T. Pharmaceuticals and personal care products in the environment: Agents of subtle change. *Environmental Health Perspectives* 1999; 107:907–938.

Díaz-Cruz MS, López de Alda MJ and Barceló D. Determination of antimicrobials in sludge from infiltration basins at two artificial recharge plants by pressurized liquid extraction–liquid chromatography–tandem mass spectrometry. *Journal of Chromatography A* 2006; 1130:72-82.

Dickert F. Molecular imprinting. *Analytical and Bioanalytical Chemistry* 2007; 389:353-354.

Dobor J, Varga M, Yao J, Chen H, Palkó G and Zára G. A new sample preparation method for determination of acidic drugs in sewage sludge applying microwave assisted solvent extraction followed by gas chromatography–mass spectrometry. *Microchemical Journal* 2009; 94:36-41.

EC, Directive 2001/82/EC of the European Parliament and the Council of 6 November 2001 on the Community code relating to vetenary medicinal products. Official Journal No. L311/67, 2001a.

EC, Directive 2001/83/EC of the European Parliament and of the Council of 6 November 2001 on the Community code relating to human medicinal products. Official Journal No. 311/1, 2001b.

Eiguren Fernández A, Sosa Ferrera Z and Santana Rodríguez JJ. Application of microwave-assisted extraction using micellar media to the determination of polychlorinated biphenyls in marine sediments. *Analytica Chimica Acta* 2001; 433:237-244.

Fent K, Weston A and Caminada D. Ecotoxicology of human pharmaceuticals. *Aquatic Toxicoly* 2006; 76:122–159.

Fontanals N, Marcé RM and Borull F. New hydrophilic materials for solid-phase extraction. *Trends in Analytical Chemistry* 2005; 24: 394-406.

Durán Guerrero E, Natera Marin R, Castro Mejias R and García Barroso C. Optimisation of stir bar sorptive extraction applied to the determination of volatile compounds in vinegars. *Journal of Chromatography A* 2006; 1104;47-53.

Göbel A, Thomsen A, McArdell CS, Alder AC, Giger W, Theiß N, Löffler D and Ternes TA. Extraction and determination of sulfonamides, macrolides, and trimethoprim in sewage sludge. *Journal of Chromatography A* 2005; 1085:179-189.

Gómez MJ, Petrovic M, Fernández-Alba AR and Barceló D. Determination of pharmaceuticals of various therapeutic classes by solid-phase extraction and liquid chromatography–tandem mass spectrometry analysis in hospital effluent wastewaters. *Journal of Chromatography A* 2006; 1114:224-233.

Gros M, Petrovic M and Barceló D. Development of a multi-residue analytical methodology based on liquid chromatography–tandem mass spectrometry (LC–MS/MS) for screening and trace level determination of pharmaceuticals in surface and wastewaters. *Talanta* 2006; 70: 678-690.

Guardabassi L, Petersen A, Olsen JE, Dalsgaard A. Antibiotic resistance in Acinetobacter spp. isolated from sewers receiving waste effluent from a hospital and a pharmaceutical plant. *Applied of Environmental Microbioly* 1998; 64:3499–502.

Halling-Sorensen B, Nielson SN, Lanzky PF, Ingerslev F, Holten Lutzhoft J and Jorgensen SE. Occurrence, fate and effects of pharmaceutical substances in the environment- A review. *Chemosphere* 1998; 35: 357-393.

Hernando MD, Heath E, Petrovic M and Barceló D. Trace-level determination of pharmaceutical residues by LC-MS/MS in natural and treated waters. A pilot-survey study. *Analytical and Bioanalytical Chemistry* 2006; 385:985-991.

Hilton MJ and Thomas KV. Determination of selected human pharmaceutical compounds in effluent and surface water samples by high-performance liquid chromatography–electrospray tandem mass spectrometry. *Journal of Chromatography A* 2003; 1015: 129-141.

Hu SG, Li L and He XW. Comparison of trimethoprim molecularly imprinted polymers in bulk and in sphere as the sorbent for solid-phase extraction and extraction of trimethoprim from human urine and pharmaceutical tablet and their determination by high-performance liquid chromatography. *Analytica Chimica Acta* 2005; 537:215-222.

Jacobsen AM and Halling-Sørensen B. Multi-component analysis of tetracyclines, sulfonamides and tylosin in swine manure by liquid chromatography-tandem mass spectrometry. *Analytical and Bioanalytical Chemistry* 2006; 384:1164-1174.

Jelic A, Petrovic M and Barceló D. Multi-residue method for trace level determination of pharmaceuticals in solid samples using pressurized liquid extraction followed by liquid chromatography/quadrupole-linear ion trap mass spectrometry. *Talanta* 2009; 80:363–371.

Juan-García A, Picó Y and Font G. Capillary electrophoresis for analyzing pesticides in fruits and vegetables using solid-phase extraction and stir-bar sorptive extraction. *Journal of Chromatography A* 2005; 1073:229-236.

Kataoka H. Automated sample preparation using in-tube solid-phase microextraction and its application - a review. *Anaytical and Bioanaytical Chemistry* 2002; 373:31–45.

Kolpin DW, Furlong ET, Meyer MT, Thurman EM, Zaugg SD, Barber LB and Buxton HT. Pharmaceuticals, hormones, and other organic wastewater contaminants in U.S. streams 1999–2000: a national reconnaissance. *Environmental Science of Technology* 2002; 36: 1202–1211.

Kümmerer K. Pharmaceuticals in the Environment, 2nd ed. Springer-Verlag 2004.

Kuster M, López de Alda MJ and Barceló D. Analysis of pesticides in water by liquid chromatography-tandem mass spectrometric techniques. *Mass Spectrometry Reviews* 2006; 25: 900-916.

Lamas JP, Salgado Petinal C, García Jares C, Llompart M, Cela R and Gómez M. Solid-phase microextraction–gas chromatography–mass spectrometry for the analysis of selective serotonin reuptake inhibitors in environmental water. *Journal of Chromatography A* 2004; 1046:241-247.

Laville N, Ait-Aissa S, Gomez E, Casellas C and Porcher JM. Effects of human pharmaceuticals on cytotoxicity, EROD activity and ROS production in fish hepatocytes. *Toxicology* 2004; 196: 41-55.

Lock CM, Chen L and Volmer DA. Rapid analysis of tetracycline antibiotics by combined solid phase microextraction/high performance liquid chromatography/mass spectrometry. *Rapid Communications in Mass Spectrometry* 1999; 13: 1744-1754.

Löffler D and Ternes TA. Determination of acidic pharmaceuticals, antibiotics and ivermectin in river sediment using liquid chromatography–tandem mass spectrometry. *Journal of Chromatography A* 2003; 1021:133-144.

Lord H and Pawliszyn J. Microextraction of drugs. *Journal of Chromatography A* 2000; 902:17-63.

Lord H and Pawliszyn J. Evolution of solid-phase microextraction technology. *Journal of Chromatography A* 2000; 885:153–193.

Lord LH, Grant RP, Walles M, Incledon B, Fahie B, Pawliszyn J. Development and evaluation of a solid-phase microextraction probe for in vivo pharmacokinetic studies. *Analytical Chemistry* 2003; 75: 5103-5115.

López de Alda MJ, Gil A, Paz E and Barceló D. Occurrence and analysis of estrogens and progestogens in river sediments by liquid chromatography-electrospray-mass spectrometry. *Analyst* 2002; 127:1299-1304.

Mahugo Santana C, Sosa Ferrera Z and Santana Rodríguez JJ. An environmentally friendly method for the extraction and determination of priority phenols in soils using microwave-assisted micellar extraction. *Analytical and Bioanalytical Chemistry* 2005; 382: 125-133.

Mahugo Santana C, Torres Padrón ME, Sosa Ferrera Z and Santana Rodríguez JJ. Development of a solid-phase microextraction method with micellar desorption for the determination of chlorophenols in water samples: Comparison with conventional solid-phase microextraction method. *Journal of Chromatography A* 2007; 1140:13–20.

McClure EL and Wong CS. Solid phase microextraction of macrolide, trimethoprim, and sulfonamide antibiotics in wastewaters. *Journal of Chromatography A* 2007; 1169:53–62.

Memoria de la Agencia española de medicamentos y productos sanitarios, 2006.

Miao X, Koenig BG and Metcalfe CD. Analysis of acidic drugs in the effluents of sewage treatment plants using liquid chromatography–electrospray ionization tandem mass spectrometry. *Journal of Chromatography A* 2002; 952:139-147.

Mitani K and Kataoka H. Determination of fluoroquinolones in environmental waters by in-tube solid-phase microextraction coupled with liquid chromatography–tandem mass spectrometry. *Analytica Chimica Acta* 2006; 562:16–22.

Montesdeoca Esponda S, Torres Padrón ME, Sosa Ferrera Z and Santana Rodríguez JJ. Solid-phase microextraction with micellar desorption and HPLC-fluorescence detection for the analysis of fluoroquinolones residues in water samples. *Analytical and Bioanalytical Chemistry* 2009; 394:927-935.

Morales S, Canosa P, Rodríguez I, Rubí E and Cela R. Microwave assisted extraction followed by gas chromatography with tandem mass spectrometry for the determination of triclosan and two related chlorophenols in sludge and sediments. *Journal of Chromatography A* 2005; 1082:128-135.

Mutavdzic D, Babic S, Asperger D, Horvat AJM and Kastelan Macan M. Comparison of different solid-phase extraction materials for sample preparation in the analysis of veterinary drugs in water samples. *Journal of Planar Chromatography* 2006; 19: 454-462.

Moeder M, Schrader S, Winkler M and Popp P. Solid-phase microextraction-gas chromatography-mass spectrometry of biologically active substances in water samples. *Journal of Chromatography A* 2000; 873:95-106.

Musteata FM and Pawliszyn J. Bioanalytical applications of solid-phase microextraction. *Trends of Analytical Chemistry* 2007; 26:36-45.

Nakamura S and Daishima S. Simultaneous determination of alkylphenols and bisphenol a in river water by stir bar sorptive extraction with in situ acetylation and thermal desorption-gas chromatography-mass spectrometry. *Journal of Chromatography A* 2004; 1038:291-294.

Nakamura S and Daishima S, Simultaneous determination of 64 pesticides in river water by stir bar sorptive extraction and thermal desorption-gas chromatography-mass spectrometry. *Analytical and Bioanalytical Chemistry* 2005; 382:99-107.

Neng NR, Pinto ML, Pires J, Marcos PM and Nogueira JMF. Development, optimisation and application of polyurethane foams as new polymeric phases for stir bar sorptive extraction. *Journal of Chromatography A* 2007; 1171: 8-14.

Nieto A, Borrull F, Marce RM and Pocurull E. Selective extraction of sulfonamides, macrolides and other pharmaceuticals from sewage sludge by pressurized liquid extraction. *Journal of Chromatography A* 2007; 1174:125–131

Okuda T, Yamashita N, Tanaka H, Matsukawa H and Tanabe K. Development of extraction method of pharmaceuticals and their occurrences found in Japanese wastewater treatment plants. *Environment International* 2009; 35:815–820.

Oliveira ARM, Cesarino EJ and Bonato PS. Solid-phase microextraction and chiral HPLC analysis of ibuprofen in urine. *Journal of Chromatography B* 2005; 818:285-291.

Oppel J, Broll G, Loffler D, Meller M, Rombke J and Ternes T. Leaching behaviour of pharmaceuticals in soil-testing-systems: A part of an environmental risk assessment for groundwater protection. *Science of Total Environment* 2004; 328:265-273.

Petrovic M, Eljarrat E, Lopez de Alda M and Barceló D. Recent advances in the mass spectrometric analysis related to endocrine disrupting compounds in aquatic environmental samples. *Journal of Chromatography A* 2002; 974:23-51.

Petrovic M, Gros M and Barceló D. Multi-residue analysis of pharmaceuticals in wastewater by ultra-performance liquid chromatography–quadrupole–time-of-flight mass spectrometry. *Journal of Chromatography A* 2006; 1124:68-81.

Pitarch E, Hernández F, ten Hove J, Meiring H, Niesing W, Dijkman E, Stolker L and Hogendoorn E. Potential of capillary-column-switching liquid chromatography–tandem mass spectrometry for the quantitative trace analysis of small molecules: Application to the on-line screening of drugs in water. *Journal of Chromatography A* 2004; 1031:1-9.

Psillakis E and Kalogenakis N. Developments in single-drop microextraction, *Trends in Analytical Chemistry* 2002; 21:54-64.

Portugal FCM, Pinto ML and Nogueira JMF. Optimization of Polyurethane Foams for Enhanced Stir Bar Sorptive Extraction of Triazinic Herbicides in Water Matrices. *Talanta* 2008; 77: 765-773.

Pozo OJ, Guerrero C,. Sancho JV, Ibañez M, Pitarch E, Hogendoorn E and Hernández F. Efficient approach for the reliable quantification and confirmation of antibiotics in water using on-line solid-phase extraction liquid chromatography/tandem mass spectrometry. *Journal of Chromatography A* 2006; 1103: 83-93.

Quintana JB, Miró M, Estela JM and Cerdá V. Automated on-line renewable solid-phase extraction-liquid chromatography exploiting multisyringe flow injection-bead injection lab-on-valve analysis. *Analytical Chemistry* 2006; 78:2832-2840.

Renew JE and Huang CH. Simultaneous determination of fluoroquinolone, sulfonamide, and trimethoprim antibiotics in wastewater using tandem solid phase extraction and liquid chromatography–electrospray mass spectrometry. *Journal of Chromatography A* 2004; 1042:113-121.

Roberts PH and Bersuder P. Analysis of OSPAR priority pharmaceuticals using high-performance liquid chromatography-electrospray ionisation tandem mass spectrometry. *Journal of Chromatography A* 2006; 1134:143-150.

Rodríguez I, Carpinteiro J, Quintana JB, Carro AM, Lorenzo RA and Cela R. Solid-phase microextraction with on-fiber derivatization for the analysis of anti-inflammatory drugs in water samples. *Journal of Chromatography A* 2004; 1024:1-8.

Rodríguez-Mozaz S, López de Alda MJ and Barceló D. Advantages and limitations of on-line solid phase extraction coupled to liquid chromatography–mass spectrometry technologies versus biosensors for monitoring of emerging contaminants in water. *Journal of Chromatography A* 2007, 1152:97-115.

Salgado-Petinal C, Lamas JP, García-Jares C, Llompart M and Cela R. Rapid screening of selective serotonin re-uptake inhibitors in urine samples using solid-phase microextraction gas chromatography-mass spectrometry. *Analytical and Bioanalytical Chemistry* 2005; 382:1351-1359.

Sanderson H, Jonson DJ, Reitsma T, Brain RA, Wilson CJ and Solomon KR. Ranking and prioritization of environmental risks of pharmaceuticals in surface waters. *Regulatory Toxicology and Pharmacology* 2004; 39: 158-183.

Silva AR, Portugal FCM and. Nogueira JMF. Advances in stir bar sorptive extraction for the determination of acidic pharmaceuticals in environmental water matrices. Comparison between polyurethane and polydimethylsiloxane polymeric phases. *Journal of Chromatography A* 2008; 1209:10–16.

Smith RM. Before the injection—modern methods of sample preparation for separation techniques. *Journal of Chromatography A* 2003; 1000:3-27.

Snow J. Solid-phase micro-extraction of drugs from biological matrices. *Journal of Chromatography A* 2000; 885:445-455.

Soliman MA, Pedersen JA and Suffet IH. Rapid gas chromatography–mass spectrometry screening method for human pharmaceuticals, hormones, antioxidants and plasticizers in water. *Journal of Chromatography A* 2004; 1029:223-227.

Sosa Ferrera Z, Padrón Sanz C, Mahugo Santana C and Santana Rodríguez JJ. The use of micellar systems in the extraction and pre-concentration of organic pollutants in environmental samples. *Trends in Analytical Chemistry* 2004; 23:469-479.

Spongberg AL and Witter JD. Pharmaceutical compounds in the wastewater process stream in Northwest Ohio. *Science of Total Environment* 2008; 397:148-157.

Stoob K, Singer HP, Stettler S, Hartmann N, Mueller SR and Stamm CH. Exhaustive extraction of sulfonamide antibiotics from aged agricultural soils using pressurized liquid extraction. *Journal of Chromatography A* 2006; 1128:1-9.

Stumpf M, Ternes TA, Wilken R, Rodrigues SV and Baumann W. Polar drug residues in sewage and natural waters in the state of Rio de Janeiro, Brazil. *Science of Total Environment* 1999; 225:135-141.

Sun C and Liu H. Application of non-ionic surfactant in the microwave-assisted extraction of alkaloids from Rhizoma Coptidis. *Analytica Chimica Acta* 2008; 612: 160-164.

Sun C, Xie Y, Tian Q and Liu H. Analysis of glycyrrhizic acid and liquiritin in liquorice root with microwave-assisted micellar extraction and pre-concentration. *Phytochemical Analysis* 2008; 19:160-163.

Tamayo FG, Turiel E and Martín-Esteban A. Molecularly imprinted polymers for solid-phase extraction and solid-phase microextraction: Recent developments and future trends. *Journal of Chromatography A* 2007; 1152:32-40.

Ternes TA. Occurrence of drugs in German sewage treatment plants and rivers. *Water Resources* 1998; 32:3245-3260.

Ternes TA, Bonerz M, Herrmann N, Löffler D, Keller E, Lacida BB and Alder AC. Determination of pharmaceuticals, iodinated contrast media and musk fragrances in sludge by LC tandem MS and GC/MS. *Journal of Chromatography A* 2005; 1067:213-223.

Theodoridis G, Koster EHM and de Jong G.J. Solid-phase microextraction for the analysis of biological samples. *Journal of Chromatography B* 2000; 745:49-82.

Theodoridis G, Lontou MA, Micholopoulos F, Sucha M and Gondova T. Study of multiple solid-phase microextraction combined off-line with high performance liquid chromatography: Application in the analysis of pharmaceuticals in urine. *Analytica Chimica Acta* 2004; 516:197-204.

Torres Padrón ME, Mahugo Santana C, Sosa Ferrera Z and Santana Rodríguez JJ. Implementation of Solid-Phase Microextraction with Micellar Desorption Method for Priority Phenolic Compound Determination in Natural Waters. *Journal of Chromatographic Science* 2008; 46:325-332.

Torres Padrón ME, Sosa Ferrera Z and Santana Rodríguez JJ. Coupling of solid-phase microextraction with micellar desorption and high performance liquid chromatography for the determination of pharmaceutical residues in environmental liquid samples. *Biomedical Chromatography* 2009; 23:1175-1185.

Turiel E, Martín-Esteban A and Tadeo JL. Multiresidue analysis of quinolones and fluoroquinolones in soil by ultrasonic-assisted extraction in small columns and HPLC-UV. *Analytica Chimica Acta* 2006; 562:30–35.

Van De Steene JC and Lambert WE. Validation of a solid-phase extraction and liquid chromatography–electrospray tandem mass spectrometric method for the determination of nine basic pharmaceuticals in wastewater and surface water samples. *Journal of Chromatography A* 2008; 1182:153–160.

Vas G and Vekey K. Solid-phase microextraction: A powerful sample preparation tool prior to mass spectrometric analysis. *Journal of Mass Spectrometry* 2004; 39: 233-254.

Vera-Candioti L, Gil García MD, Martínez Galera M and Goicoechea HC. Chemometric assisted solid-phase microextraction for the determination of anti-inflammatory and antiepileptic drugs in river water by liquid chromatography–diode array detection. *Journal of Chromatography A* 2008; 1211:22-32.

Wen Y, Zhou BS, Xu Y, Jin SW and Feng, YQ. Analysis of estrogens in environmental waters using polymer monolith in-polyether ether ketone tube solid-phase microextraction combined with high-performance liquid chromatography. *Journal of Chromatography A* 2006; 1133:21–28.

Witte W. Medical consequences of antibiotics use in agriculture. *Science* 1998; 279: 996-997.

Yang JJ and Metcalfe CD. Fate of synthetic musks in a domestic wastewater treatment plant and in an agricultural field amended with biosolids. *Science of Total Environment* 2006; 363:149-165.

Zorita S, Barri T and Mathasson L. A novel hollow-fibre microporous membrane liquid–liquid extraction for determination of free 4-isobutylacetophenone concentration at ultra trace level in environmental aqueous samples. *Journal of Chromatography A* 2007; 1157:30-37.

In: Biomedical Chromatography
Editors: John T. Elwood, pp. 141-159
ISBN: 978-1-60741-291-5

Chapter 5

ANALYSIS OF CATECHOLAMINES WITH PEROXYOXALATE CHEMILUMINESCENCE REACTION DETECTION

Makoto Tsunoda
Laboratory of Bio-Analytical Chemistry, Graduate School of Pharmaceutical Sciences, The University of Tokyo, 7-3-1, Hongo, Bunkyo-ku, Tokyo 113-0033, Japan

ABSTRACT

Peroxyoxalate chemiluminescence (POCL) reaction is known as a useful, selective and sensitive detection system especially when coupled with separation techniques such as high-performance liquid chromatography (HPLC) or capillary electrophoresis. We have developed analytical methods for catecholamines and/or their metabolites using HPLC-POCL reaction detection. This article reviews 1) the analytical methods for catecholamines, 2) the analytical methods for catecholamines and their 3-*O*-methyl metabolites, and 3) the application of these methods to bio-samples for the clarification of the role of catecholamines metabolism in blood pressure regulation.

1. INTRODUCTION

Catecholamines (CAs) include compounds with a dihydroxyphenyl group and an amine group. Three catecholamine compounds, norepinephrine (NE), epinephrine (E), and dopamine (DA), are known to be synthesized *in vivo*.

The biosynthetic pathway and the chemical structures of the CAs are shown in Figure 1. Tyrosine hydroxylase, which converts tyrosine to L-3,4-dihydroxyphenylalanine (L-DOPA), is the rate-limiting step in the production of CAs. Dopamine is produced from L-DOPA by dopa decarboxylase. NE and E are also produced enzymatically. As shown in Figure 1, CAs are metabolized and inactivated by two enzymes, monoamine oxidase and catechol-*O*-methyltransferase (COMT). Homovanillic acid (HVA) and vanillylmandelic acid (VMA) are the final metabolites. CAs participate in a wide variety of *in-vivo* regulation systems involving adrenergic and dopaminergic receptors.

Recent developments in analytical methods have enabled us to determine CAs and/or their metabolites in biological fluids with high sensitivity and precision [1]. The availability of such methods has prompted clarification of the role of CAs in several pathological conditions. Quantitation of CA concentrations in plasma and/or urine is also useful as an index for several diseases, for example hypertension [2], pheochromocytoma [3], and neuroblastoma [4]. High-performance liquid chromatography (HPLC) with electrochemical detection is most often used for analysis of CAs and their metabolites, because of its high sensitivity and selectivity. Fluorescence detection is also frequently used for analysis of CAs, which have native fluorescence. This fluorescence is not, however, suitable for quantitation of traces of CAs in real samples, because the emission wavelengths are short (approx. 320 nm) and it has been proved they are obstructed by those of endogenous compounds in real samples. Fluorescence derivatization of CAs combined with the use of separation techniques such as HPLC or capillary electrophoresis (CE) has therefore become quite common. Mass spectrometry (MS) has recently become a popular method because of its structural specificity. In the early stages of investigation, gas chromatography (GC) was used, coupled with MS for the identification of CAs and their metabolites [5, 6]. LC or CE coupled with MS (LC-MS or CE-MS) is currently a powerful technique because of recent instrumental developments, including the interface. Thus, LC–MS and CE–MS have been adopted in several recent studies of CAs [7].

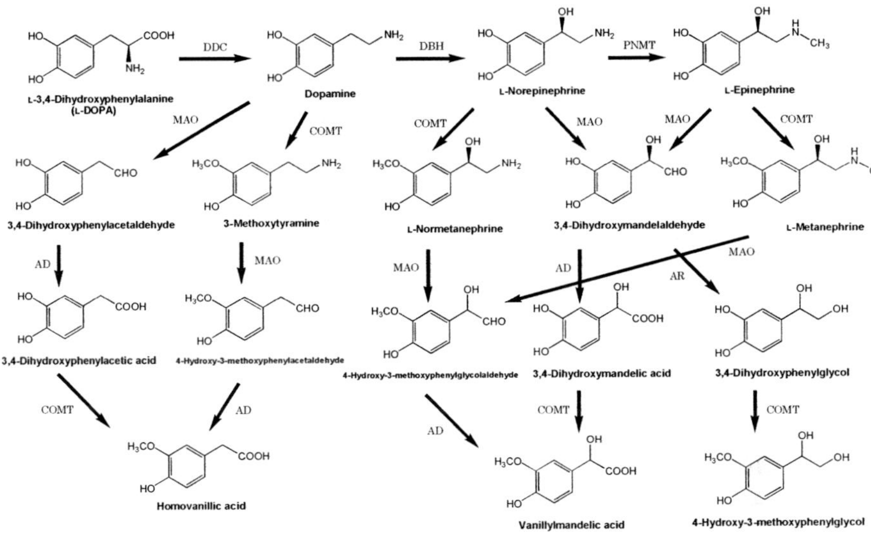

Figure 1. Pathway of catecholamine biosynthesis and metabolism. *DDC*, Dopa decarboxylase; *DBH*, dopamine β-hydroxylase; *PNMT*, phenylethanolamine N-methyltransferase; *COMT*, catechol-*O*-methyltransferase; *MAO*, monoamine oxidase; *AD*, aldehyde dehydrogenase; *AR*, aldehyde reductase.

In this chapter, I focus on chemiluminescence detection. The chemiluminescence detection is known to be more sensitive and selective than fluorescence detection. Thus, peroxyoxalate chemiluminescence (POCL) and electro-generated chemiluminescence are used for the determination of CAs.

POCL was first introduced by Chandross in 1963 [8]. He found a bright chemiluminescence when the reaction was carried out by adding the acid chloride to hydrogen peroxide containing a fluorescent compound (fluorophore) such as anthracene, 9,10-diphenylanthracene or *N*-methylacridone.

POCL reaction involves hydrogen peroxide oxidation of an aryl oxalate ester in the presence of fluorophore. Therefore, it can be used for the detection of a fluorophore or hydrogen peroxide. Since the advent of its use with combination of HPLC [9], POCL reaction coupled with flow injection analysis (FIA), HPLC or CE has been extensively studied for the detection and quantitation of fluorophores or fluorescent derivatized compounds. It has been used also to detect hydrogen peroxide as well as the analytes following their conversion to hydrogen peroxide by enzymatic reaction. In POCL reaction, the elimination of an exciting light source, which is used in the fluorescence detection, reduces background emission or light source instability, and the sensitivity thus improves to a 10–100 fold over the fluorescence detection.

We have developed analytical methods for CAs and/or their metabolites using HPLC-POCL reaction detection. This article reviews 1) the analytical methods for CAs, 2) the analytical methods for CAs and their 3-*O*-methyl metabolites, and 3) the application of these methods to bio-samples for the clarification of the role of CAs metabolism in blood pressure regulation.

2. Analytical Methods for Catecholamines with HPLC-POCL Detection [10]

2.1. High Selectivity of the Method

As mentioned above, CAs include compounds with a dihydroxyphenyl group and an amine group. To selectively detect CAs, a cation exchange column, as a precolumn, was first used to adsorb only amines including CAs. This was then fluorescently derivatized with ethylenediamine that is known to specifically react with catechol moiety. This strategy is illustrated in Figure 2.

2.2. Procedures of the Method

Figure 3 shows the block diagram of the column-switching HPLC system for the automated analyzer of CAs, which includes the following steps,

1. A sample injected from the auto sampler is introduced to the precolumn (cation exchange column). Amines including CAs are adsorbed in the precolumn.
2. CAs are desorbed from the precolumn, and are separated on an ODS column.
3. Separated CAs are derivatized with ethylenediamine, which produces fluorescent compounds.
4. Fluorescent derivatized CAs are reacted with chemiluminescence reagent (bis [4-nitro-2-(3,6,9-trioxadecyloxycarbonyl)phenyl]oxalate (TDPO) and hydrogen peroxide), and are detected.

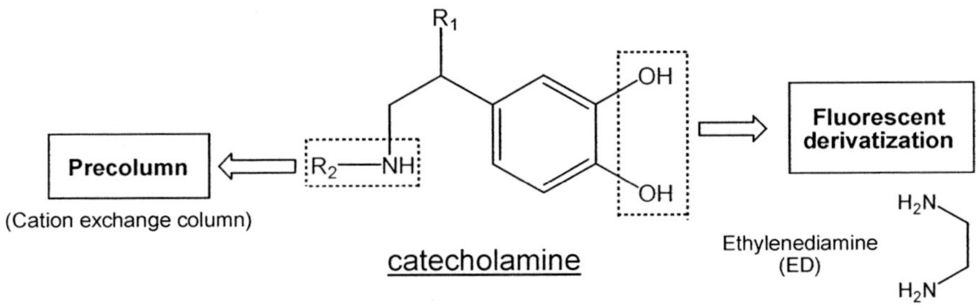

R_1	R_2	
OH	H	Norepinephrine (NE)
OH	CH_3	Epinephrine (E)
H	H	Dopamine (DA)

Figure 2. Selectivity of the analytical method for catecholamines.

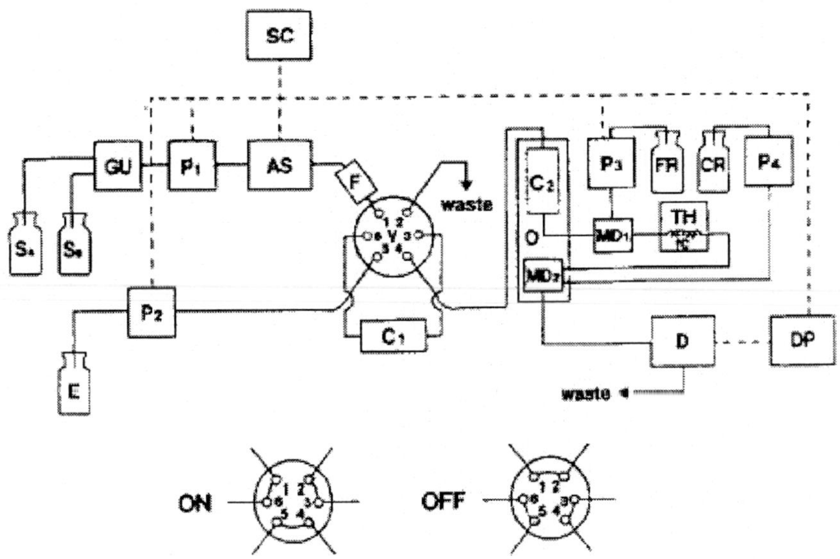

Figure 3. Block diagram of the column-switching HPLC system for the automated analyzer of catecholamines. The system comprised: S_A and S_B, solvent A and solvent B; E, eluent; FR, fluorogenic reagent solution; CR, chemiluminogenic reagent solution; GU, gradient unit; P_1, P_2, P_3, and P_4, HPLC pumps; AS, autosampler; SC, system controller; V, high-pressure switching-valve; F, linefilter; C_1, precolumn; C_2, analytical column; MD_1 and MD_2, rotating flow mixing devices; TH, thermostatically controlled bath; O, column oven; rc, reaction coil; D, chemiluminescence detector; DP, data processor. Reprinted from ref [10] with permission from Wiley.

2.3. Chromatogram of Standard Catecholamines and Validation of the Method

Figure 4 shows a chromatogram of a mixture of standard CAs, containing 100 fmol each of NE, E, DA and 200 fmol of *N*-methyldopamine (*N*-MeDA) as internal standard. The peaks were well resolved in the chromatogram. The lowest detection limits were 1, 0.7 and 1 fmol for NE, E and DA, respectively, at an signal to noise ratio of 2.

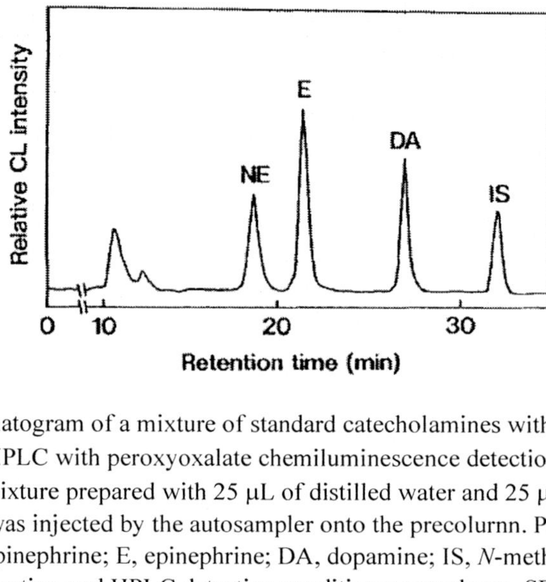

Figure 4. Chromatogram of a mixture of standard catecholamines with in-line extraction and HPLC with peroxyoxalate chemiluminescence detection. A 50 μL volume of the mixture prepared with 25 μL of distilled water and 25 μL of the sample dilution buffer was injected by the autosampler onto the precolurnn. Peaks (100 fmol each): NE, norepinephrine; E, epinephrine; DA, dopamine; IS, *N*-methyldopamine (200 fmol). Extraction and HPLC detection conditions: precolumn, SERUMOUT-CEX; solvent A, 10 mM potassium phosphate buffer (pH 7.5)/ethanol (92 + 8, v/v) (1 mL/min); adsorption time in the precolumn, 2 min; eluent, 75 mM potassium acetate buffer (pH 3.2)/50 mM potassium phosphate buffer (pH 3.2)/acetonitrile (92.15+4.85+3, v/v/v) containing 1 mM sodium hexanesulphonate (0.5 ml/min); desorption time from the precolumn, 2 min.; solvent B, phosphoric acid/water/acetonitrile (2+48+50, v/v/v) (1 mL/min); precolumn clean-up time, 5 min; column, TSK gel ODS-8OTs 250 × 4.6 mm i.d.; column temperature, 40 °C; fluorogenic reagent solution, 105 mM ethylenediamine and 175 mM imidazole in acetonitrile/ethanol (90+10, v/v) (0.32 mL/min); reaction coil for fluorogenic reaction, 15 m × 0.5 mm i.d.; reaction temperature, 80 °C; chemiluminogenic reaction solution, 0.25 mM TDPO, 150 mM hydrogen peroxide and 110 mM TFA in dioxane/ethyl acetate (50+50, v/v) (1.4 mL/min). Reprinted from ref [10] with permission from Wiley.

Table 1. Precision of the present method

	Rat plasma			Human plasma		
	fmol/mL	SD	%	fmol/ml	SD	%
Within-day assay ($n = 3$)						
NE	921	±28.6	3.1	3662	±102	2.8
E	661	±22.6	3.4	472	±12.6	2.7
DA	158	±5.2	3.3	122	±3.8	3.1
Between-day assay ($n = 3$)						
NE	990	±40.6	4.1	3708	±185	5.0
E	630	±23.2	3.7	490	±18.6	3.8
DA	180	±8.8	4.9	110	±4.4	4.0

Reprinted from ref [10] with permission from Wiley.

The calibration curve for each CA showed linearity in the range of 5.0 fmol to 500 fmol. The coefficients of correlation were 0.999 or bigger for each CA.

The recovery experiment was conducted for CAs spiked in 50 μL of human or rat plasma (50.0, 100, 250, 350 and 500 fmol). *N*-MeDA was used as internal standard. The recoveries for NE, E and DA were 97 ± 3.5%, 92 ± 4.0% and 103 ± 4.0% in human plasma ($n = 3$) and 103 ± 4.0%, 99.5 ± 2.5% and 100 ± 3.0% in rat plasma ($n = 3$).

Within-day and between-day precision for NE, E and DA are briefly summarized in Table 1. The relative standard deviations obtained using 50 μL of human or rat plasma were less than 4% in case of within-day assay and less than 5% in case of between-day assay. The data suggest that the precision of the proposed method is appropriate for routine assay of CAs in human and rat plasma.

2.4. Concentration of Catecholamines in Human and Rat Plasma

Chromatograms of CAs in human and rat plasma are shown in Figure 5 (a) and (b). The plasma concentration of NE, E and DA in a healthy human were 3670, 490 and 110 fmol/mL, respectively, which were within a normal range of CAs: NE 2030 ± 1080; E 390 ± 220; DA 70 ± 60 fmol/mL [11] or NE 4200 ± 1700; E 1000 ± 300 fmol/mL [12]. The values for CAs in rat plasma were 960, 3040 and 180 fmol/mL, respectively.

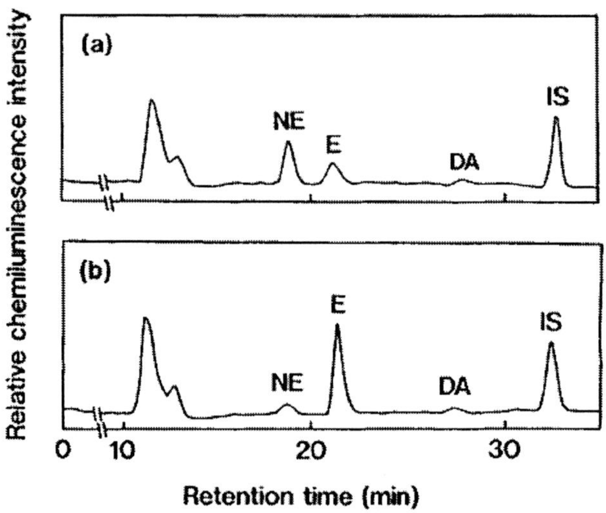

Figure 5. Chromatograms obtained from (a) human plasma, and (b) Sprague-Dawley rat plasma by the analytical method for catecholamines. Samples of 50 μL (25 μL of plasma + 25 μL of the dilution buffer) were injected by the autosampler onto the precolumn. Reprinted from ref [10] with permission from Wiley.

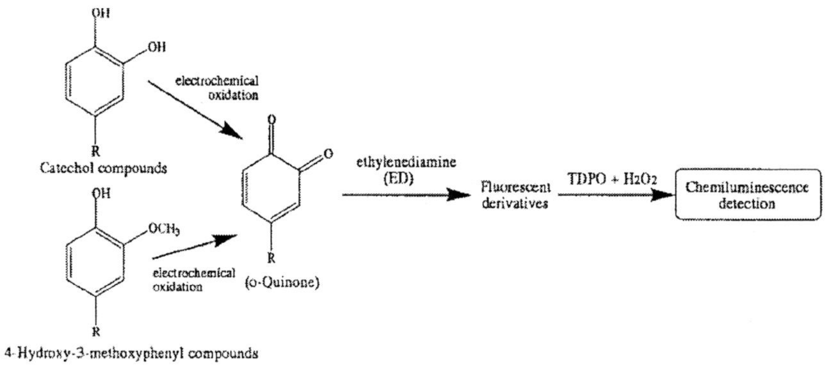

Figure 6. Electrochemical oxidation of catechol and 4-hydroxy-3-methoxyphenyl compounds followed by ethylenediamine reaction and chemiluminescence detection. Reprinted from ref [17] with permission from Elsevier.

One of the most important advantages of the present method is that the volume of plasma required is very small (25-50 μL) compared to bigger volumes in other methods previously reported: 100-200 μL [13]; 500-750 μL [14]; 1 mL [15]; 2 mL [16].

The method overcomes problems derived from time consuming sample pretreatment. This system will be very useful for routine assay of plasma NE, E, DA and synthetic CAs in clinical laboratories. It will also be useful for assays in biological samples, which involve small volumes of samples being tested, as the sensitivity of the present method is sufficient enough to determine CAs in 25 μL of human plasma or rat plasma.

3. Analytical Methods for Catecholamines and Their 3-*O*-Methyl Metabolites [17]

As mentioned in Introduction section, one of the major metabolic pathways for CAs is the *O*-methylation at their 3-hydroxyl groups by COMT, transforming NE, E, and DA to normetanephrine (NMN), metanephrine (MN), and 3-methoxytyramine (3-MT), respectively [18]. Since COMT is localized in extraneuronal tissues, the various changes of *O*-methylated metabolites of CAs in plasma provide the information of extraneuronal inactivation of CAs. Therefore, simultaneous determination of CAs and their 3-*O*-methyl metabolites in plasma is useful for examination of sympathoadrenal activity as well as the metabolizing activity of CAs in extraneuronal tissues.

In this section, a simultaneous analytical method for CAs and their 3-*O*-methyl metabolites is described, which included on-line extraction and separation in a similar way to analytical method for CAs, and then electrochemically oxidized them to their respective *o*-quinones by a coulometric reactor. Further, fluorogenic derivatization and peroxyoxalate chemiluminescence reaction were carried out in the same way as mentioned above.

3.1. Coulometric Oxidation

As shown in Figure 6, in the present method, prior to the fluorescence derivatization reaction with ethylenediamine, 4-hydroxy-3-methoxyphenyl compounds must be converted into the respective *o*-quinone compounds by employing the electrochemical coulometric detector to oxidize these compounds. The effects of the applied potential of the coulometric detector on the conversion reaction were evaluated by measurement of the fluorescence intensity of the fluorescent products of CAs and their 3-*O*-methyl metabolites.

As shown in Figure 7, the fluorescence intensities for 3-*O*-methyl metabolites were proportional to the applied potential in the range 0.1–0.3 V (versus a H_2/H^+ reference electrode), whereas for CAs their intensities were almost the same within this range. However, the fluorescence intensities of CAs and their 3-*O*-methyl metabolites decreased when the potentials were higher than 0.35 V. Hence, a potential of 0.3 V was selected for the simultaneous measurement of CAs and their 3-*O*-methyl metabolites.

3.2. Validation of the Method and Application to Rat Plasma

Calibration curve, accuracy, and precision of the method

The calibration curves showed linearity in the range of 10 to 500 fmol for NE, E, DA, NMN, and MN, and 20 to 500 fmol for 3-MT. The coefficients of correlation were 0.988 or higher for all CAs and their 3-*O*-methyl metabolites. The limits of detection (signal-to-noise ratio of 3) were about 3 fmol for NE, E, and DA, 5 fmol for NMN, and 10 fmol for MN and 3-MT. These detection limits for NE, E, and DA were similar to those mentioned above [10]. The sensitivity of 3-*O*-methyl metabolites was better than the methods previously reported (60 fmol for NMN and MN, 70 fmol for 3-MT [12], 20 fmol for NMN, 25 fmol for MN, 42 fmol for 3-MT [19], 25 fmol for NMN, and 50 fmol for MN [20]). The accuracy and the precision for NE, E, DA, NMN, and MN were shown in Table 2. The data suggest that the proposed method is appropriate for the routine assay of CAs and their 3-*O*-methyl metabolites in 50 μL of rat plasma.

Table 2.

Accuracy and Precision of the Present Method

	Accuracy (%) (n = 3)	Intraday assay CV (%) (n = 5)	Interday assay CV (%) (n = 5)
NE	97	2.6	7.7
E	92	2.5	9.1
DA	86	7.6	8.6
NMN	87	2.7	6.3
MN	91	6.8	7.2

Reprinted from ref [17] with permission from Elsevier.

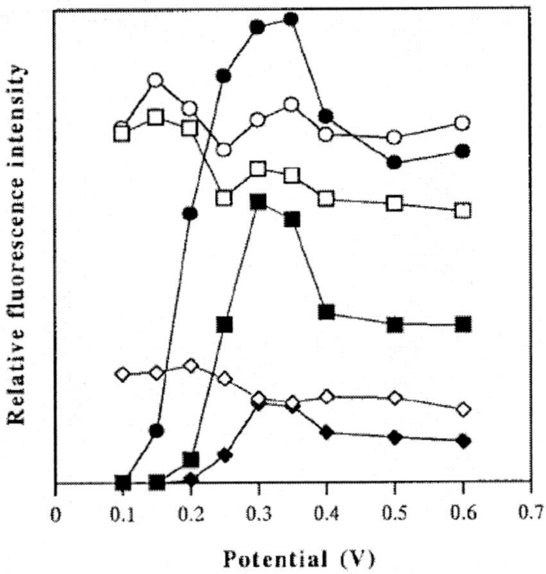

Figure 7. Effect of applied potential on the fluorescence peak areas for catecholamines and their 3-*O*-methyl metabolites. One hundred picomoles each of NE (□), E (□), DA (○), NMN (■), MN (□), and 3-MT (●) was injected. HPLC conditions: eluent, 75 mM potassium acetate buffer (pH 3.2)/50 mM potassium phosphate buffer (pH 3.2)/acetonitrile (90.25/4.75/5, v/v/v) containing 4 mM sodium 1-hexane sulfonate, at a flow rate of 0.5 ml/min; fluorogenic reagent, 105 mM ethylenediamine and 175 mM imidazole in acetonitrile/ethanol (90/10, v/v), at a flow rate of 0.32 ml/min; reaction coil, 0.5 mm i.d. × 10 m; and reaction temperature, 80°C. The fluorescence intensity was monitored at 505 nm with excitation at 430 nm. Reprinted from ref [17] with permission from Elsevier.

Determination of catecholamines and their 3-O-methyl metabolites in rat plasma

Figure 8 shows a typical chromatogram obtained from a Sprague-Dawley rat plasma sample (50 μL). With the exception of 3-MT, the peaks for NE, E, DA, NMN, and MN in the rat plasma were clearly shown with no interference from endogenous compounds. This implies the high selectivity of the proposed method, since it includes on-line selective extraction of amines by the precolumn and a selective ethylenediamine condensation for catechol compounds. On the other hand, the detection with electrochemical oxidation is selective only for catechol compounds [20-23]. In the chromatogram of rat plasma sample without 4-methoxytyramine (4-MT, internal standard), no peak appeared at the retention time of 4-MT (data not shown), suggesting that 4-MT was suitable for use as an internal standard in rat plasma. The concentrations of CAs and their 3-*O*-methyl metabolites in rat plasma ($n = 3$) for NE, E, DA, NMN, and MN were 1.05 ± 0.03, 0.64 ± 0.02, 0.19 ± 0.01, 0.51 ± 0.02, and 0.26 ± 0.01 pmol/ml, respectively. These values were similar to the reported values (NE, E, and DA: 0.92 – 0.96, 0.63 – 0.66, and 0.16 – 0.18 pmol/ml [10]; NE, E, NMN, and MN: 1.20 ± 0.14, 0.57 ± 0.18, 0.46 ± 0.05, and 0.17 ± 0.03 pmol/ml [24]; and NE, E, NMN, and MN: 0.99 ± 0.07, 0.75 ± 0.07, 0.41 ± 0.02, and 0.19 ± 0.02 pmol/ml [25]).

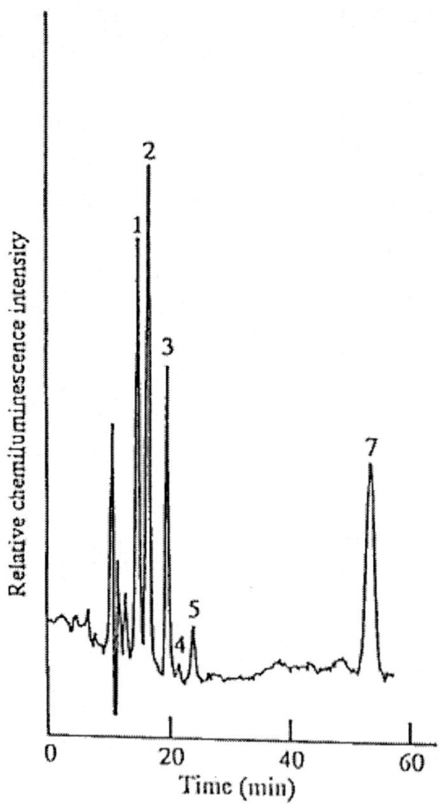

Figure 8. Chromatogram obtained from a Sprague–Dawley (SD) rat plasma by the proposed method. A sample of 150 μL (50 μL of plasma + 100 μL of the dilution buffer) was injected by the autosampler onto the precolumn. Peaks: 1, NE; 2, E; 3, NMN; 4, DA; 5, MN; 7, 4-MT (internal standard). Reprinted from ref [17] with permission from Elsevier.

4. Application of These Methods to Bio-Samples for the Clarification of the Role of Catecholamines Metabolism in Blood Pressure Regulation

When the blood pressure is reduced, the first reflex by the homeostatic system is via the sympathetic barorefrex to induce a release of neurotransmitter NE from the nerve endings. The release is directly related to the degree of the acute hypotension: the greater the hypotensive effect, the greater the amount of NE is released [26, 27]. Previously, we demonstrated that the ratio of the increase of plasma NE concentration to the reduction of blood pressure was attenuated in spontaneously hypertensive rats (SHR) as compared to the age-matched normotensive Wistar-Kyoto (WKY) rats [28-30].

From this finding, we posed a fundamental question: Is the metabolism of NE different in the two strains? To solve this, benidipine, a calcium antagonist, was administered to SHR and WKY rats to induce a release of NE via the sympathetic baroreflex as mentioned above. Then, the released NE is extraneuronally inactivated to its *O*-methyl metabolite, NMN, by COMT. Since the ratio of plasma NMN to plasma NE was thought to be a parameter for the metabolic inactivation, we simultaneously measured plasma NE and NMN concentration using the developed method as mentioned above.

In an acute hypotension, when the blood pressure was reduced by benidipine, the ratio of NMN to NE in plasma was lower in SHR than in WKY rats (Figure 9). These data indicate that, in an induced hypotension where NE rapidly increases, the extraneuronal inactivation of the released NE to NMN is decreased. This suggests that the activity of COMT is attenuated in SHR as compared to WKY rats.

Next, we attempted to search for the main COMT(s) that inactivate(s) released NE during acute hypotension. Since COMTs exist in most tissues in a soluble form (S-COMT) as well as a membrane-bound form (MB-COMT), it is important to examine whether one of these subtypes plays a more crucial role than the other in the inactivation of released NE.

To measure the activity of COMT in various tissues of the rat, we developed methods for COMT activity measurement using CAs as substrate [31-35]. Using the developed method, we determined S- and MB-COMT enzyme activities and the amount of protein in two representative rat tissues, the liver and the kidney, as well as in erythrocytes, and found that both the activites and the amounts of MB-COMT in the liver were lower in SHR than in WKY rats [36]. This result indicate that liver MB-COMT may be a relevant factor in blood pressure regulation in rats.

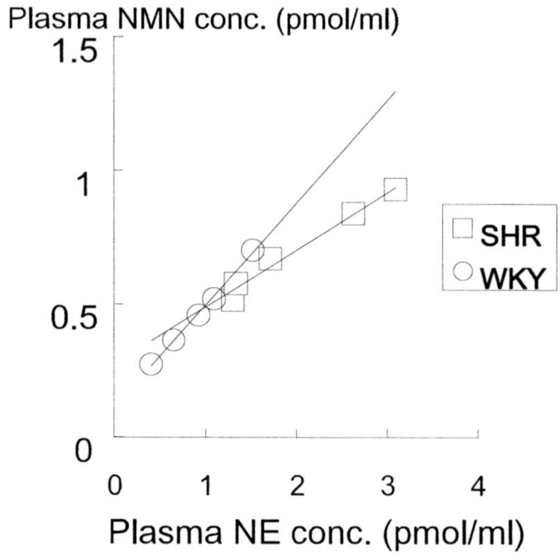

Figure 9. Relationship between plasma NE and NMN concentration in the face of an acute hypotension. The slope for WKY rats was significantly greater than that for SHR ($p<0.05$).

5. Conclusion

This chapter reviewed the analytical methods for the determination of catecholamines (norepinephrine, epinephrine and dopamine), and catecholamines and their 3-*O*-methyl metabolites with HPLC-peroxyoxalate chemiluminescence reaction detection. The developed method was very sensitive and selective for the determination of catecholamines and/or their 3-*O*-methyl metabolites, and clarified the role of catecholamines metabolism in blood pressure regulation.

REFERENCES

[1] Tsunoda, M. (2006). Recent advances in methods for the analysis of catecholamines and their metabolites. *Analytical and Bioanalytical Chemistry*, *386*, 506-514.

[2] Goldstein, D. S. (1983). Plasma-Catecholamines and Essential-Hypertension - an Analytical Review. *Hypertension*, *5*, 86-99.

[3] Goldstein, D. S., Eisenhofer, G., Flynn, J. A., Wand, G. & Pacak, K. (2004). Diagnosis and localization of pheochromocytoma. *Hypertension*, *43*, 907-910.

[4] Hyland, K. (2008). Clinical utility of monoamine neurotransmitter metabolite analysis in cerebrospinal fluid. *Clinical Chemistry*, *54*, 633-641.

[5] Canfell, C., Binder, S. R. & Khayambashi, H. (1982). Quantitation of Urinary Normetanephrine and Metanephrine by Reversed-Phase Extraction and Mass-Fragmentographic Analysis. *Clinical Chemistry*, *28*, 25-28.

[6] Yi, Z. & Brown, P. R. (1991). Chromatographic Methods for the Analysis of Basic Neurotransmitters and Their Acidic Metabolites. *Biomedical Chromatography*, *5*, 101-107.

[7] Bergquist, J., Sciubisz, A., Kaczor, A. & Silberring, J. (2002). Catecholamines and methods for their identification and quantitation in biological tissues and fluids. *Journal of Neuroscience Methods*, *113*, 1-13.

[8] Chandross, E. A. (1963). A New Chemiluminescent System. *Tetrahedron Letters*, 761-765.

[9] Kobayashi, S. & Imai, K. (1980). Rotating Flow Mixing Device for Post Column Reaction in High-Performance Liquid-Chromatography. *Analytical Chemistry*, *52*, 1548-1549.

[10] Prados, P., Higashidate, S. & Imai, K. (1994). A Fully Automated Hplc Method for the Determination of Catecholamines in Biological Samples Utilizing Ethylenediamine Condensation and Peroxyoxalate Chemiluminescence Detection. *Biomedical Chromatography*, *8*, 1-8.

[11] Mitsui, A., Nohta, H. & Ohkura, Y. (1985). High-Performance Liquid-Chromatography of Plasma-Catecholamines Using 1,2-Diphenylethylenediamine as Precolumn Fluorescence Derivatization Reagent. *Journal of Chromatography*, *344*, 61-70.

[12] Nohta, H., Yamaguchi, E., Ohkura, Y. & Watanabe, H. (1989). Measurement of Catecholamines, Their Precursor and Metabolites in

Human-Urine and Plasma by Solid-Phase Extraction Followed by High-Performance Liquid-Chromatography with Fluorescence Derivatization. *Journal of Chromatography-Biomedical Applications*, *493*, 15-26.

[13] Higashidate, S. & Imai, K. (1992). Determination of Femtomole Concentrations of Catecholamines by High-Performance Liquid-Chromatography with Peroxyoxalate Chemiluminescence Detection. *Analyst*, *117*, 1863-1868.

[14] Yamatodani, A. & Wada, H. (1981). Automated-Analysis for Plasma Epinephrine and Norepinephrine by Liquid-Chromatography, Including a Sample Cleanup Procedure. *Clinical Chemistry*, *27*, 1983-1987.

[15] Kamahori, M., Taki, M., Watanabe, Y. & Miura, J. (1991). Analysis of Plasma-Catecholamines by High-Performance Liquid-Chromatography with Fluorescence Detection - Simple Sample Preparation for Precolumn Fluorescence Derivatization. *Journal of Chromatography-Biomedical Applications*, *567*, 351-358.

[16] Dejong, J., Point, A. J. F., Tjaden, U. R., Beeksma, S. & Kraak, J. C. (1987). Determination of Catecholamines in Urine (and Plasma) by Liquid-Chromatography after Online Sample Pretreatment on Small Alumina or Dihydroxyborylsilica Columns. *Journal of Chromatography-Biomedical Applications*, *414*, 285-300.

[17] Tsunoda, M., Takezawa, K., Santa, T. & Imai, K. (1999). Simultaneous automatic determination of catecholamines and their 3-O-methyl metabolites in rat plasma by high-performance liquid chromatography using peroxyoxalate chemiluminescence reaction. *Analytical Biochemistry*, *269*, 386-392.

[18] Mannisto, P. T. & Kaakkola, S. (1999). Catechol-O-methyltransferase (COMT): Biochemistry, molecular biology, pharmacology, and clinical efficacy of the new selective COMT inhibitors. *Pharmacological Reviews*, *51*, 593-628.

[19] Jeon, H. K., Nohta, H., Nagaoka, H. & Ohkura, Y. (1991). Simultaneous Determination of Catecholamine-Related Compounds by High-Performance Liquid-Chromatography with Postcolumn Chemical Oxidation Followed by a Fluorescence Reaction. *Analytical Sciences*, *7*, 257-262.

[20] Lenders, J. W. M., Eisenhofer, G., Armando, I., Keiser, H. R., Goldstein, D. S. & Kopin, I. J. (1993). Determination of Metanephrines in Plasma by Liquid-Chromatography with Electrochemical Detection. *Clinical Chemistry*, *39*, 97-103.

[21] Cao, G. M. & Hoshino, T. (1996). High-performance liquid chromatographic determination of catecholamines, serotonin and their metabolites with three-potential electrochemical detection. *Analytical Sciences*, *12*, 183-188.

[22] Mashige, F., Matsushima, Y., Miyata, C., Yamada, R., Kanazawa, H., Sakuma, I., Takai, N., Shinozuka, N., Ohkubo, A. & Nakahara, K. (1995). Simultaneous Determination of Catecholamines, Their Basic Metabolites and Serotonin in Urine by High-Performance Liquid-Chromatography Using a Mixed-Mode Column and an 8-Channel Electrochemical Detector. *Biomedical Chromatography*, *9*, 221-225.

[23] Volin, P. (1992). Determination of Urinary Normetanephrine, Metanephrine and 3-Methoxytyramine by High-Performance Liquid-Chromatography with Electrochemical Detection - Comparison between Automated Column-Switching and Manual Dual-Column Sample Purification Methods. *Journal of Chromatography-Biomedical Applications*, *578*, 165-174.

[24] Lenders, J. W. M., Kvetnansky, R., Pacak, K., Goldstein, D. S., Kopin, I. J. & Eisenhofer, G. (1993). Extraneuronal Metabolism of Endogenous and Exogenous Norepinephrine and Epinephrine in Rats. *Journal of Pharmacology and Experimental Therapeutics*, *266*, 288-293.

[25] Eisenhofer, G. & Finberg, J. P. M. (1994). Different Metabolism of Norepinephrine and Epinephrine by Catechol-O-Methyltransferase and Monoamine-Oxidase in Rats. *Journal of Pharmacology and Experimental Therapeutics*, *268*, 1242-1251.

[26] Imai, K., Higashidate, S., Prados, P. R., Santa, T., Adachi-Akahane, S. & Nagao, T. (1994). Relation between Blood-Pressure and Plasma-Catecholamine Concentration after Administration of Calcium-Antagonists to Rats. *Biological & Pharmaceutical Bulletin*, *17*, 907-910.

[27] Higashidate, S., Imai, K., Prados, P., Adachi-Akahane, S. & Nagao, T. (1994). Relations between Blood-Pressure and Plasma Norepinephrine Concentrations after Administration of Diltiazem to Rats - Hplc-Peroxyoxalate Chemiluminescence Determination on an Individual Basis. *Biomedical Chromatography*, *8*, 19-21.

[28] Prados, P., Santa, T., Fukushima, T., Homma, H., Kasai, C., Martin, M. A., del Castillo, B. & Imai, K. (1998). Age-related weakening of baroreflex-mediated sympathetic activity in spontaneously hypertensive rats in response to blood pressure reduction. *Hypertension Research-Clinical and Experimental*, *21*, 147-153.

[29] Prados, P., Santa, T., Homma, H., Doi, H., Narita, H., del Castillo, B., Martin, M. A. & Imai, K. (1997). Comparison of the sympathetic nervous system activity between spontaneously hypertensive and Wistar-Kyoto rats to respond to blood pressure reduction. *Biological & Pharmaceutical Bulletin, 20*, 341-344.

[30] Tsunoda, M., Takezawa, K., Santa, T., Ina, Y., Nagashima, K., Ohmori, K., Kobayashi, S. & Imai, K. (2000). New approach for measurement of sympathetic nervous abnormality in conscious, spontaneously hypertensive rats. *Japanese Journal of Pharmacology, 83*, 39-45.

[31] Tsunoda, M., Takezawa, K. & Imai, K. (2001). A method for the measurement of catechol-O-methyltransferase activity using norepinephrine, an endogenous substrate. *Analyst, 126*, 637-640.

[32] Tsunoda, M., Takezawa, K., Masuda, M. & Imai, K. (2002). Rat liver and kidney catechol-O-methyltransferase activity measured by high-performance liquid chromatography with fluorescence detection. *Biomedical Chromatography, 16*, 536-541.

[33] Masuda, M., Tsunoda, M. & Imai, K. (2003). High-performance liquid chromatography-fluorescent assay of catechol-O-methyltransferase activity in rat brain. *Analytical and Bioanalytical Chemistry, 376*, 1069-1073.

[34] Masuda, M., Tsunoda, M., Yusa, Y., Yamada, S. & Imai, K. (2002). Assay of catechol-O-methyltransferase activity in human erythrocytes using norepinephrine as a natural substrate. *Annals of Clinical Biochemistry, 39*, 589-594.

[35] Hirano, Y., Tsunoda, M., Funatsu, T. & Imai, K. (2005). Rapid assay for catechol-O-methyltransferase activity by high-performance liquid chromatography-fluorescence detection. *Journal of Chromatography B-Analytical Technologies in the Biomedical and Life Sciences, 819*, 41-46.

[36] Tsunoda, M., Tenhunen, J., Tilgmann, C., Arai, H. & Imai, K. (2003). Reduced membrane-bound catechol-O-methyltransferase in the liver of spontaneously hypertensive rats. *Hypertension Research, 26*, 923-927.

In: Biomedical Chromatography
Editors: John T. Elwood, pp. 141-159
ISBN: 978-1-60741-291-5

Chapter 6

PHOTOPRODUCTS OF NAPROXEN IN ALCOHOLIC SOLVENTS

Hsin-Tsung Ho [a,b], An-Bang Wu [c*], Pen-Yuan Lin [c], Jen-Ai Lee [c] and Pen-Yueh Hung [c]

[a] Department of Laboratory Medicine, Mackay Memorial Hospital, Taipei 10449, Taiwan , R.O.C.
[b] Mackay Medicine, Nursing and Management College, Taipei 11272, Taiwan, R.O.C.
[c] College of Pharmacy, Taipei Medical University, Taipei 11031, Taiwan, R.O.C.

ABSTRACT

A sample of 5 mM naproxen in methanol or ethanol was photo-irradiated with a Hanovia 200 W high-pressure quartz Hg lamp. In total, 6 photoproducts derived from each sample were observed from the HPLC chromatograms. Four major photoproducts were separated, and their structures elucidated by EI-MS and various spectroscopic methods. A reaction scheme of naproxen in alcoholic solvents is proposed: the photochemical reaction routes occur mainly via decarboxylation and

* Corresponding author: E-mail: anbangwu@tmu.edu.tw, College of Pharmacy, Taipei Medical University, 250 Wu Hsing Street, Taipei 11031, Taiwan, ROC, Tel: 886-2-27361661 ext. 6121. Fax and direct phone: 886-2-27366518.

esterification, followed by oxidation with singlet oxygen to produce an alcohol and a ketone.

Key words: Naproxen; HPLC; EI-MS; Photochemistry.

INTRODUCTION

Several years after naproxen was first synthesized by Harrison et al. [1], the U.S. Food and Drug Administration (FDA) in 1976 approved the marketing of the drug under the trade names of Aleve, Naprosyn, and others. Since then, naproxen, 2-(6-methoxy-2-naphthyl)propionic acid, or its sodium salt has been widely used as a non-steroidal anti-inflammatory drug (NSAID) which has analgesic and antipyretic activities [2]. However, on December 20, 2004, the FDA released a statement warning users that according to a recent study, preliminary information showed some evidence of increased risk of cardiovascular events to patients taking naproxen. In addition, for a long time, the clinical photosensitivity also seemed to be the most commonly reported adverse cutaneous reactions including wheal-and-flare [3] after the administration of naproxen [4,5]. A general survey of the mechanisms of photosensitization induced by drugs was reported by Quintero and Miranda [6]. More recently, drug-induced cutaneous photosensitivity was presented by Moore [7] who stated that the photochemical and photobiological mechanisms underlying the adverse reactions caused by NSAIDs are mainly free radical in nature, but reactive oxygen species are also involved. The photochemical mechanisms for NSAIDs that contain the 2-aryl propionic acid group, e.g., naproxen, involve decarboxylation as the primary step, with subsequent free radical activity. In aerated systems, the reactive excited singlet form of oxygen is produced with high efficiency. This form of oxygen is highly reactive towards lipids and proteins. In a photodegradation study of naproxen in aqueous phosphate-buffered solutions, Castell et al. [8] found that the photomixtures obtained in the presence of oxygen were clearly more toxic to cultured hepatocytes than those obtained under anaerobic conditions. They explained that the observed toxicity can be attributed to the presence of drug-derived peroxidic species.

Previous studies concerning naproxen mainly centered on its physiological effects. The intermediates or photoproducts seemed to be linked to the cause of the phototoxicity [9-13]. In the present study, we attempted to focus on the isolation and identification of the major photoproducts after photolysis of naproxen in methanol and ethanol solutions (as medically modified) by high-performance liquid chromatography (HPLC). The structural identification of the major photoproducts was determined using various spectroscopic methods [14].

EXPERIMENTAL

General Methods and Materials

Melting points were determined on a Büchi (Zurich, Switzerland) B-540 apparatus. Elemental microanalysis was performed at National Taiwan or Tatung University using a Heraeus Vario EL-III model (Hanau, Germany). (S)-Naproxen was purchased from Sigma Chemical (St. Louis, MO, USA). Liquid chromatographic (LC) grade methanol and guaranteed reagent (GR grade) D-chloroform were from Merck (Darmstadt, Germany). GR grade glacial acetic acid and absolute ethanol were the products of Ridel-deHaën (Seelze, Germany). LC grade acetonitrile was purchased from Labscan Asia (Bangkok, Thailand). The chemicals were used as received. The purities of the chemicals were checked by running HPLC using blank samples.

HPLC Apparatus and Assay Conditions

An Alcott (Norcross, GA, USA) 760 HPLC pump system equipped with a Linear (Reno, NV, USA) UVIS-206 detector, a DATAAPEX Clarity (Prague, Czech Republic) chromatography station, CSW 1.7 integrator, and a GL Sciences (Tokyo, Japan) preparative Inertsil ODS-3 column of 250 × 10-mm i.d. was used with a mobile phase of CH_3CN-CH_3OH-H2O (deionized water containing 1% acetic acid) (40: 20: 40, v/v/v). The UV detector was set at 230 nm. A manual Hamilton (Bonaduz, Switzerland) 80565 injector was used with the flow rate controlled to 5 mL/min. The volume of each injection was 200 μL.

Photochemical Reactor and Irradiation Conditions

A Panchum (Taipei, Taiwan) PR-2000 reactor, which was equipped with sixteen 8-W low-pressure quartz mercury lamps (Sankyo Denki G8T5E) as the light source, was used. Irradiation was performed with the samples in stoppered quartz tubes mounted vertically on a merry-go-round rack at a speed of 6 rpm. The light intensity of the monochromatic radiation was measured at 306 nm to be 3.25 mW/cm^2 using a UVX (UVP, Upland, CA, USA) digital radiometer (serial No. E. 16768).

Various Spectrometers

For NMR, a Brüker (Ettlingen, Germany) ACE-500 FT-NMR (500 MHz) spectrometer was used. All samples including naproxen and the photoproducts were prepared by their dissolution in $CDCl_3$ to concentrations of about 10 mg/mL. Distortionless enhancement by polarization transfer (DEPT) was adopted to distinguish quaternary carbons. 2D NMR of heteronuclear multiple quantum coherence (HMQC) for determining the ^1J (C, H) correlation, and heteronuclear multiple bond connectivity (HMBC) for showing the ^2J (C, H) and ^3J (C, H) long-range coupling relations were conducted. Chemical shifts, δ, in ppm of ^1H and ^{13}C NMR spectra were measured with respect to $CHCl_3$ (7.26 and 77.00 ppm for ^1H and ^{13}C, respectively).

For IR, a Bio-Rad Digilab (Cambridge, MA, USA) FTS-40 FT-IR was used. Each sample was mixed with KBr in a 1: 100 (w/w) ratio to make the disc for taking the IR spectrum.

For UV, a Heλions Alpha (Unicam Instrument, Cambridge, UK) was used. Each sample of approximately 100 μM in methanol was prepared and placed in a quartz cell for measurement of the UV spectrum.

For EI-MS, a Hewlett-Packard 5989B gas chromatograph-mass spectrometer (*Palo Alto*, *CA,* USA) was used.

Sample Preparation and Photodegradation of Naproxen

An amount of 115 mg (5 mM) of naproxen was accurately weighed and placed in a 100-mL volumetric flask. Methanol or ethanol was slowly added to make the concentration of the sample exactly 500 μM. Each sample was transferred to a quartz tube. The two tubes were stoppered and irradiated in a photochemical reactor for 7 days. Samples were filtered using a Millipore membrane (0.45 μm thickness), and the filtrate was then subjected to HPLC separation. The major photoproducts of the irradiated mixture were separately collected using a preparative HPLC. Four fractions of the major photoproducts, namely **1, 2, 3,** and **4** (in methanol) and **1, 2, 3,** and **5** (in ethanol) were collected. The solvents were evaporated and then subjected to a series of spectroscopic analyses.

Characterization of Naproxen and Its Photoproducts by Spectroscopic Methods

Naproxen (NAP), 6-methoxy-α-methyl-2-naphthaleneacetic acid

M.P. 152~154°C; UV, λmax in nm (absorbance): 231 (1.884), 262 (0.153), 271 (0.152), 332 (0.046); IR (KBr) in cm^{-1}: 3199 (b, νO-H), 1728 (s, νC=O), 1604 (m, naphthyl,νC=C); ^1H NMR (in $CDCl_3$, δin ppm relative to $CHCl_3$): 7.690 (d, 1H, C1), 7.417 (d, 1H, J = 9.59 Hz, C3), 7.700 (d, 1H, J = 9.95 Hz, C4), 7.111 (d, 1H, C5), 7.143 (dd, 1H, J = 11.06 Hz, C7), 7.699 (d, 1H, J = 9.95 Hz, C8), 3.914 (s, 3H, C9), 3.878 (q, 1H, J = 7.02 Hz, C10), 1.597 (d, 3H, J = 6.99 Hz, C11). ^{13}C NMR (in $CDCl_3$): 180.610 (C12), 157.701 (C6), 128.800 (C8'), 134.835 (C2), 126.163 (C1), 133.807 (C4'), 127.200 (C4), 129.280 (C8), 126.125(C3), 119.012 (C7), 105.593 (C5), 55.278 (C9), 45.226 (C10), 18.102 (C11); EI-MS (70 ev): *m/z* (rel. int. %) 230 (61), 185 (100), 170 (26), 141 (41), 115 (40). Anal. calcd for C14H14O3 (230.26): C, 73.02%, H, 6.12%; Found: C, 73.62%, H, 6.16%.

6-Methoxy-α-methyl-2-naphthalenemethanol (1)

M.P. 110~112°C; UV, λmax in nm (absorbance): 209 (0.294), 230 (0.994), 331 (0.023); IR (KBr) in cm-1: 3336 (b, νO-H), 1607 (m, naphthyl,νC=C); 1H NMR (in CDCl3): δ7.744 (d, 1H, J = 6.28 Hz, C8), 7.696 (d, 1H, C1), 7.482 (d, 1H, J = 8.20 Hz, C2), 7.781 (d, 1H, J = 8.38 Hz, C4), 7.122 (d, 1H, C5), 7.169 (d, 1H, J = 7.60 Hz, C7), 3.945 (s, 3H, C9), 3.929 (q, 1H, J = 6.35 Hz, C10), 1.479 (d, 3H, J = 6.46 Hz, C11). 13C NMR (in CDCl3): δ157.625 (C6), 139.186 (C2), 127.139 (C8'), 134.348 (C4'), 125.144 (C1), 126.884 (C4), 129.322 (C8), 124.823(C3), 118.683 (C7), 105.723 (C5), 70.470 (C10), 55.326 (C9), 24.573 (C11); EI-MS (70 ev): m/z (rel. int. %) 202

(74), 187 (88), 159 (76), 144 (100), 128 (34), 115 (93). Anal. calcd for C13H14O2 (202.25): C, 77.20%, H, 6.98%; Found: C, 77.39%, H, 6.88%.

2-Acetyl-6-methoxynaphthalene (2)

M.P. 107~109°C; UV, λmax in nm (absorbance): 209 (0.812), 226 (0.889), 241 (1.807), 310 (0.651); IR (KBr) in cm^{-1}: 1675 (s, νC=O), 1621 (m, naphthyl,νC=C); ^{1}H NMR (in $CDCl_3$): δ7.792 (d, 1H, C1), 7.708 (d, 1H, *J* = 8.54 Hz, C2), 7.971 (d, 1H, *J* = 7.95 Hz, C4), 7.971 (d, 1H, *J* = 9.27 Hz, C8), 7.101 (d, 1H, C5), 7.167 (dd, 1H, *J* = 9.03 Hz, C7), 3.900 (s, 3H, C9), 2.654 (s, 3H, C11). ^{13}C NMR (in $CDCl_3$): δ197.633 (C10), 159.599 (C6), 132.448 (C4'), 137.120 (C2), 130.944 (C8), 126.921 (C1), 129.766 (C8'), 127.649(C4), 124.468 (C3), 119.531 (C7), 105.618 (C5), 55.233 (C9), 26.348 (C11); EI-MS (70 ev): *m/z* (rel. int. %) 200 (51), 185 (100), 157 (52), 142 (29), 114 (48). Anal. calcd for C13H12O2 (200.24): C, 77.98%, H, 6.04%; Found: C, 77.85%, H, 6.04%.

2-Ethyl-6-methoxynaphthalene (3)

M.P. 58.2~59.6°C; UV, λmax in nm (absorbance): 229 (1.449), 261 (0.092), 270 (0.092), 318 (0.031), 332, (0.035); IR (KBr) in cm-1: 1605 (m, naphthyl,νC=C); 1H NMR (in CDCl3): δ7.571 (d, 1H, C1), 7.697 (d, 1H, J = 3.84 Hz, C8), 7.129 (d, 1H, C5), 7.330 (d, 1H, J = 8.19 Hz, C3), 7.680 (d, 1H, J = 8.57 Hz, C4), 7.149 (d, 1H, J = 2.30 Hz, C7), 3.925 (s, 3H, C9), 2.800 (q, 2H, J = 7.53 Hz, C10), 1.339 (t, 3H, J = 7.29 Hz, C11). 13C NMR (in CDCl3): δ157.072 (C6), 139.428 (C2), 132.886 (C4'), 126.665 (C1), 125.397 (C3), 129.165 (C8), 127.444 (C4), 118.556 (C7), 128.873 (C8'), 105.673 (C5), 55.252 (C9), 28.807 (C10), 15.586 (C11); EI-MS (70 ev): m/z (rel. int. %) 186 (70), 171 (100), 141 (14), 128 (69), 115 (28). Anal. calcd for C13H14O (186.25): C, 83.84%, H, 7.58%; Found: C, 83.77%, H, 7.62%.

2-(6-Methoxy-2-naphthalenyl)propionic acid, methyl ester (4)

M.P. 94~95°C; UV, λmax in nm (absorbance): 211 (0.289), 232 (1.21), 262 (0.088), 271 (0.084), 317 (0.024); IR (KBr) in cm^{-1}: 1739 (s, νC=O), 1605 (m, naphthyl,νC=C); ^{1}H NMR (in $CDCl_3$): δ7.698 (d, 1H, *J* = 8.50 Hz, C8), 7.666 (d, 1H, C1), 7.406 (q, 1H, *J* = 8.09 Hz, C3), 7.715 (d, 1H, *J* = 8.50 Hz, C4), 7.116 (d, 1H, C5), 7.144 (dd, 1H, *J* = 8.87 Hz, C7), 3.910 (s, 3H, C9), 3.863 (q, 1H, *J* = 7.15 Hz, C10), 3.674 (s, 3H, C13), 1.583 (d, 3H, *J* = 7.19 Hz, C11). ^{13}C NMR (in $CDCl_3$): δ175.106 (C12), 157.629 (C6), 128.911 (C8'), 135.652 (C2), 126.157 (C1), 133.675 (C4'), 127.144 (C4), 129.239 (C8), 125.906(C3), 118.957 (C7), 105.584 (C5), 55.278 (C9), 52.000 (C13), 45.325

(C10), 18.559 (C11); EI-MS (70 ev): *m/z* (rel. int. %) 244 (51), 185 (100), 170 (17), 141 (22), 115 (14). Anal. calcd for C15H16O3 (244.29): C, 73.84%, H, 6.61%; Found: C, 73.78%, H, 6.64%.

2-(6-Methoxy-2-naphthalenyl)propionic acid, ethyl ester (5)

M.P. 80~82°C; UV,λmax in nm (absorbance): 232 (1.476), 262 (0.111), 272 (0.111), 332 (0.040); IR (KBr) in cm-1: 1742 (s, νC=O), 1613 (m, naphthyl,νC=C); 1H NMR (in CDCl3): δ7.698 (d, 1H, J = 8.91 Hz, C8), 7.673 (d, 1H, C1), 7.418 (q, 1H, J = 8.50 Hz, C3), 7.716 (d, 1H, J = 8.83 Hz, C4), 7.120 (d, 1H, C5), 7.144 (dd, 1H, J = 8.83 Hz, C7), 3.915 (s, 3H, C9), 3.842 (q, 1H, J = 7.14 Hz, C10), 4.1112 (q, 2H, J = 7.25 Hz, C13), 1.577 (d, 3H, J = 7.25 Hz, C11), 1.208 (t, 3H, J = 7.17 Hz, C14). 13C NMR (in CDCl3): δ174.670 (C12), 157.601 (C6), 128.928 (C8'), 135.841 (C2), 126.232 (C1), 133.645 (C4'), 127.467 (C4), 129.256 (C8), 125.887(C3), 118.904 (C7), 105.588 (C5), 55.286 (C9), 60.723 (C13), 45.480 (C10), 18.586 (C11), 14.116(C14); EI-MS (70 ev): m/z (rel. int. %) 258 (41), 185 (100), 170 (19), 141 (32), 115 (27). Anal. calcd for C16 H18O3 (258.31): C, 74.48%, H, 7.03%; Found: C, 74.40%, H, 7.05%.

RESULTS AND DISCUSSION

HPLC Separation of the Photoproducts of Naproxen

When it comes to the study of the photochemical behavior of NSAIDs, naproxen represents one of the simplest model drugs for investigation of photoproducts. With a stable naphthyl nucleus possessing 61 Kcal mol^{-1} of resonance energy, the photolytic reactions always occur at the side chain of the propionic acid. In this study, 5 mM of naproxen in methanol or ethanol was prepared by exposure to a normal atmosphere. Samples were placed in a photochemical reactor and were subjected to irradiation for 7 days. After the HPLC analyses, 4 major photoproducts were observed from the chromatogram in methanol (shown in Figure 1) or ethanol, respectively. Their retention times in methanol are arranged in increasing order of **1**, 11.35; naproxen (**NAP**), 14.11; **2**, 17.43; **3**, 31.63; and **4**, 47.27 min; in ethanol, they were **1**, 10.43; naproxen (**NAP**), 12.97; **2**, 14.79; **3**, 20.67; and **5**, 24.81 min (Table 1).

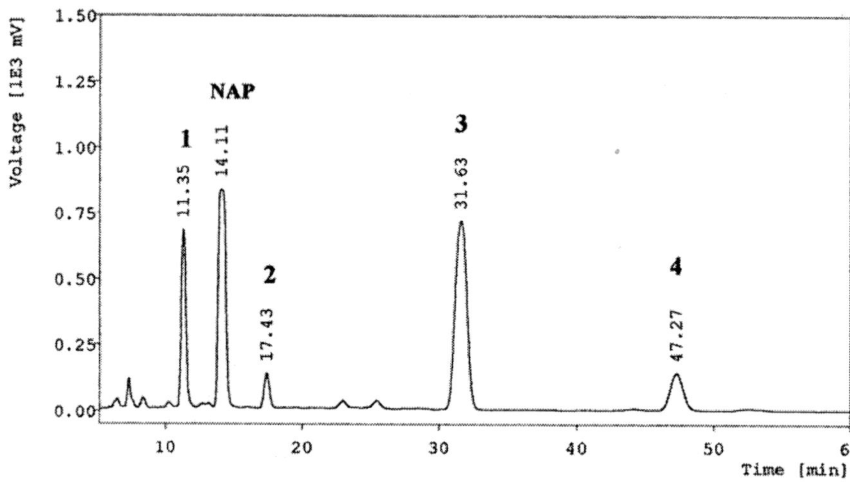

Figure 1. HPLC chromatogram of naproxen in methanol using an Inertsil 5 ODS-3V column (250 × 4.6 mm i.d.) with the mobile phase of CH_3CN-CH_3OH-1% HOAc in deionized H_2O = 40:20:40, v/v/v. The UV detector set at 230 nm.

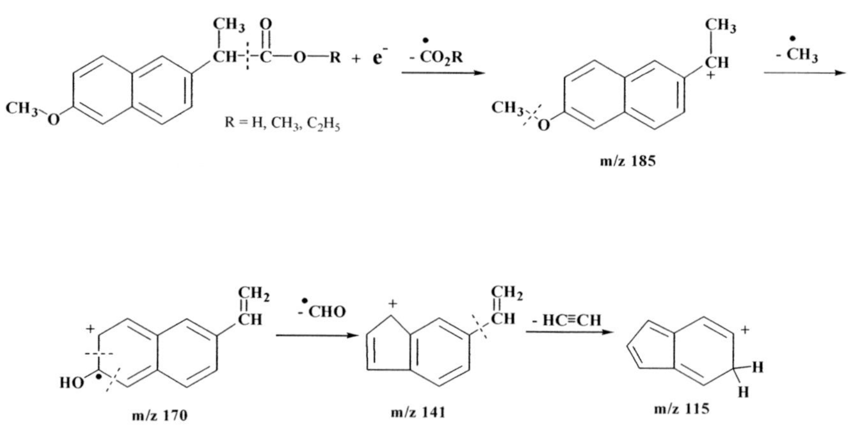

Figure 2. EI-MS fragmentation pattern of **NAP**, **4**, and **5**.

Structural Characterization of the Photoproducts

Based on EI-MS of Nap, 4, and 5, the typical fragmentation patterns of the molecular ions and the subsequent fragments are essentially similar as listed in Figure 2. The structural features of the photoproducts are thus inferred and also tabulated in Table 1. All signals of the ^{1}H, ^{13}C, and 2D NMR spectra of photoproducts were correlated in detail and properly assigned. IR spectra with the absorption bands of characteristic functional groups were very well correlated with the proposed structures, and are reported in Experimental Section. By the characteristics of the above various spectroscopic results, the structures of photoproducts 1, 2, 3, 4, and 5 can be identified as 6-methoxy-α-methyl-2-naphthylmethanol, 2-acetyl-6-methoxynaphthalene, 2-ethyl-6-methoxynaphthalene, 2-(6-methoxy-2-naphthalenyl)propionic acid, methyl ester, and 2-(6-methoxy-2-naphthalenyl)propionic acid, ethyl ester, respectively.

Table 1. The photoproducts derived from naproxen in alcoholic solvents

Compound	Retention Time (min)		Molecular Weight (g mol^{-1})	Chemical Structure[c]
1	11.35[a]	10.43[b]	202	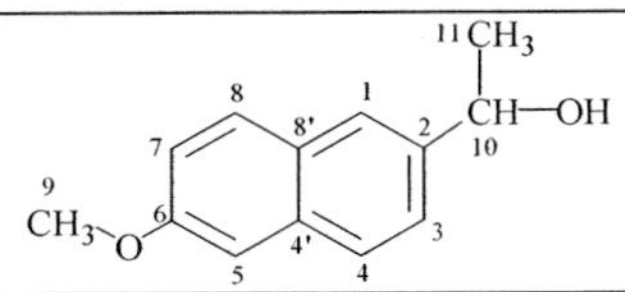
NAP	14.11	12.97	230	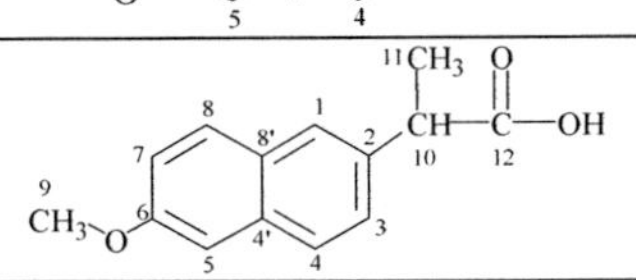
2	17.43	14.79	200	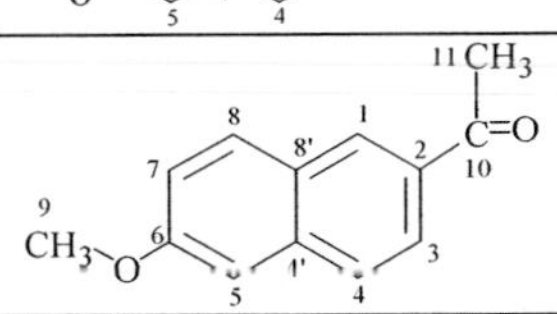
3	31.63	20.67	186	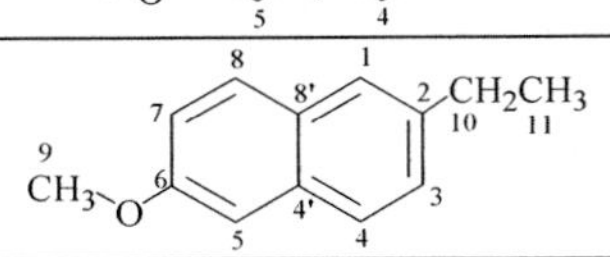
4	47.27		244	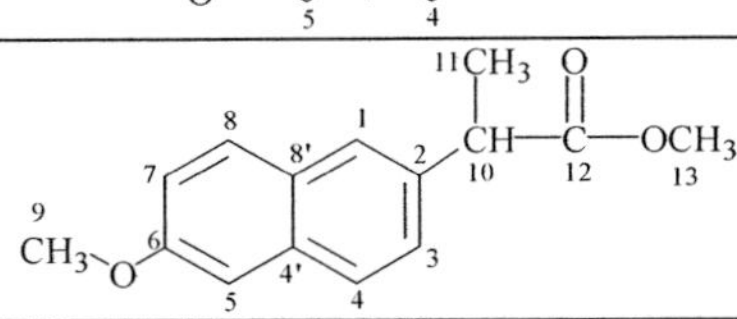
5		24.81	258	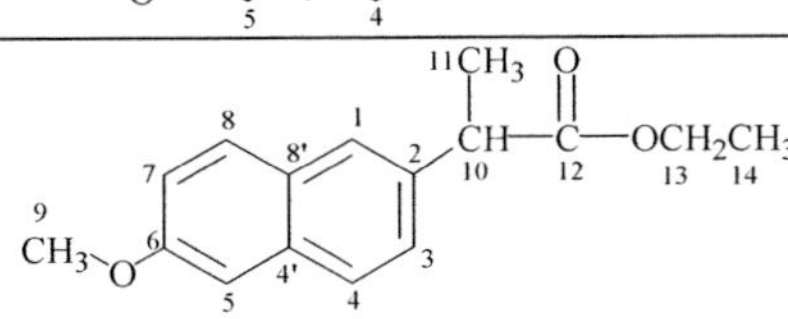

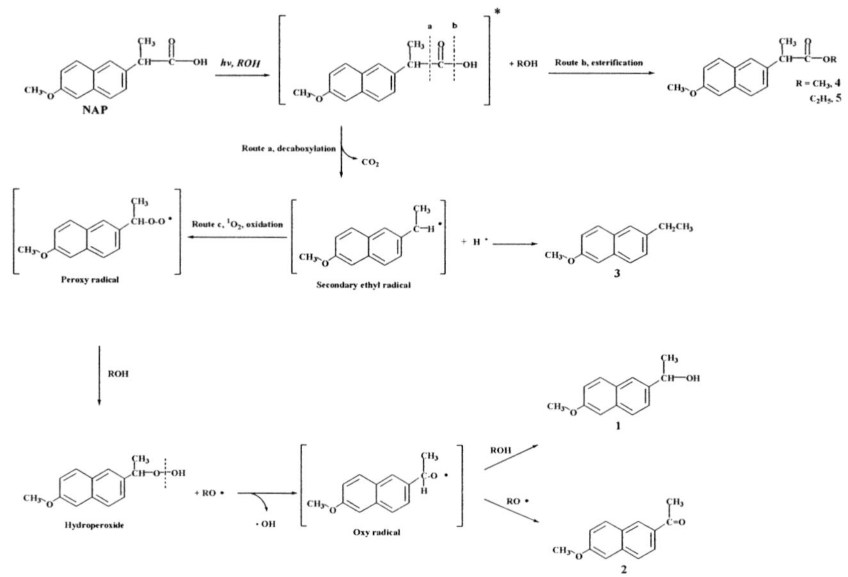

Figure 3. A photodegradation reaction scheme of naproxen (a) decarboxylation; (b) esterification; and (c) oxidation with singlet oxygen

A Proposed Reaction Scheme for Naproxen

When naproxen in an aqueous PBS solution is exposed to light, the major reaction routes are simply two ways of decarboxylation [9] and oxidation [10], which excludes the possibilities of photodechlorination [15], α-cleavage of a ketone [16] and other reactions. Moore and Chappuis [10] reported that in an aqueous solution, naproxen proceeds via decarboxylation by photoirradiation with a 125 W medium-pressure mercury lamp. Constanzo et al. [11] extended the knowledge in greater detail that the irradiation of naproxen in a deaerated PBS solution underwent a decarboxylation process via intermediate radicals, while under aerobic conditions photo-oxidation occurred. A molecular mechanism involving free radicals and singlet oxygen as important intermediates was proposed. Bosca et al. [9] found that the decarboxylation of naproxen and its sodium salt was achieved by means of chemical and electrochemical oxidation. Castell et al. [8] carried out the photodegradation of naproxen in an aqueous buffered solution and reported that the main photoproducts were decarboxylated ethyl, oxidized 1-hydroxyethyl, and acetyl side chains. By a close examination in the present study, photoproducts with similar reaction pattern were observed to be derived from naproxen in alcoholic (organic) solvents. A proposed photolytic reaction scheme for naproxen is shown in Figure 3.

CONCLUSION

In this study, we found that under a normal atmosphere, the photolysis of naproxen in methanol or ethanol solution with similar reaction pattern leads to the formation of photoproducts of decarboxylation and esterification to become a methyl or an ethyl ester, and oxidation to an alcohol and a ketone.

ACKNOWLEDGMENT

This study was sponsored by the Mackay Memorial Hospital (93MMH-TMU-15) of Taipei, Taiwan, ROC.

REFERENCES

[1] Harrison, IT; Lewis, B; Nelson, P; Rooks, W; Roszkowski, A; Tomolonis, A; Fried, JH. Nonsteroidal anti-inflammatory agents. I. 6-substituted-2-naphthylacetic acid. *J. Med. Chem.*, 1970, 13, 203-205.

[2] Brogden, RN; Heel, RC; Speight, TM; Avery, GS. Naproxen up to date - review of its pharmacological properties and therapeutic efficacy and use in rheumatic disease and pain states. *Drugs*, 1979, 18(4), 241-277.

[3] Kaidbey, KH; Mitchell, FN. Photosensitizing potential of certain nonsteroidal anti-inflammatory agents. *Arch. Dermatol*, 1989, 125, 783-786.

[4] Shelley, WB; Elpern, DJ; Shelley, ED. Naproxen photosensitization demonstrated by challenge. *Curtis*, 1986, 38,169-170.

[5] Levy, ML; Barron, KS; Eichenfield, A; Honig, PJ. Naproxen-induced pseudoporphyria: a distinctive photodermatitis. *J. Pediatr*, 1990, 117, 660-664.

[6] Quintero, B; Miranda, MA. Mechanisms of photosentitization induced by drugs: a general survey. *Ars Pharm*, 2000, 41, 27-46.

[7] Moore, DE. Drug-induced cutaneous photosensitivity. *Drug Safety*,

2002, 25(5), 345-372.

[8] Castell, JV; Gomez-Lechon, MJ; Grassa, C; Martinez, LA; Miranda, MA; Tarrega, P. Involvement of drug-derived peroxides in the phototoxicity of naproxen and tiaprofenic acid. *Photochem. Photobiol*, 1993, 57, 486-490.

[9] Bosca, F; Martinez-Marnez, R; Miranda, MA; Primo, J; Soto, J; Vano, L. Oxidative decarboxylation of naproxen. *J. Pharm. Sci.*, 1992, 81, 479-482.

[10] Moore, DE; Chappuis, PP. A comparative study of the photochemistry of the non-steroidal anti-inflammatory drugs, naproxen, benoxaprofen and indomethacin. *Photochem. Photobiol*, 1988, 47, 173-180.

[11] Constanzo, LL; De Guidi, G; Condorelli, G..; Cambria A; Fama, M. Molecular mechanism of naproxen photosensitization in red blood cells. *J. Photochem. Photobiol. B. Biol.*, 1989, 3, 223-235.

[12] Komuro, M; Nagatsu, Y; Higuchi, T; Hirobe, M. Oxidative decarboxylation of carboxylic acids by iron prophyrin-iodosylbenzene system. *Tetrahedr. Lett*, 1992, 33, 4949-4952.

[13] Becker, L; Eberlein-König, B; Przybilla, B. Phototoxicity of non-steroidal anti-inflammatory drugs: in vitro studies with visible light. *Acta Derm. Venereol*, 1996, 76(5), 337-340.

[14] Ho, HT; Liou, YB; Lin, PY; Wang, PY; Lin, DZL; Wu AB. Photolysis of NSAIDs. V. Photoproducts of naproxen in alcoholic solvents. Sixth Tetrahedron Symposium. *Challenges in Organic Chemistry,* Bordeaux, France, poster abstract No, 2005, 58, 140.

[15] Sheu, MT; Ho, HO; Wang, PY; Liou, YP; Wu, AB. Photolysis of NSAIDs. I. Photodegradation products of carprofen determined by LC-ESI-MS. *J. Chromatogr. Sci.*, 2003, 41, 200-204.

[16] Wang, CC; Chen, F.; Chen, CJ; Chao, SH; Wu, AB. Photolysis of NSAIDs. IV. Photoproducts of Zomepirac Determined by LC-ESI-MS. *Biomed. Chromatogr*, 2004, 18, 820-825.

INDEX

C

D

H

I

K

L

M

N

O

P

Q

R

S